HIGH-TECH CAREERS
FOR
LOW-TECH PEOPLE

HIGH-TECH CAREERS FOR LOW-TECH PEOPLE

2ND EDITION

William A. Schaffer

TEN SPEED PRESS
BERKELEY, CALIFORNIA

Copyright © 1999 by William A. Schaffer

All rights reserved. No part of this book may be reproduced in any form, except brief excerpts for the purpose of review, without written permission of the publisher.

A Kirsty Melville Book

Ten Speed Press
P.O. Box 7123
Berkeley, California 94707
www.tenspeed.com

Distributed in Australia by Simon and Schuster Australia, in Canada by Ten Speed Press Canada, in New Zealand by Southern Publishers Group, in South Africa by Real Books, in Southeast Asia by Berkeley Books, and in the United Kingdom and Europe by Airlift Book Company.

Cover design by Pentagram, San Francisco
Interior design by Lisa Patrizio
Library of Congress Cataloging-in-Publication Data
Schaffer, William A.
    High-tech careers for low-tech people / by William A. Schaffer.
       p. cm.
    ISBN 1-58008-039-1
    1. Industrial technicians—Vocational guidance. 2. High technology—Vocational guidance. I. Title.
    TA158.S33 1999
    004'.023—dc21   98-53532
            CIP

First printing, 1999
Printed in Canada
3 4 5 6 7 8 9 10 - 04 03 02 01 00

CONTENTS

ACKNOWLEDGMENTS vi

INTRODUCTION. viii

1 — Can You Have a Career in High Tech? 1

2 — What Is "High Tech"? 12

3 — The Jobs 18

4 — Starting Your High-Tech Career 81

5 — Computers in Thirty Minutes 131

6 — The High-Tech Culture 164

7 — High-Tech Careers, and How to ~~Survive~~ Thrive in Them 188

8 — Start-ups 208

9 — International Work in High Tech 218

10 — Final Words of Wisdom 246

WORKS CITED 250

INDEX. 251

ACKNOWLEDGMENTS

To Harry, Paul, and Amy—my favorite examples
of low-tech people who are making a difference
in the high-tech industry.

This book would never have seen the light of day without the enthusiastic participation and support of many people. I can't list all of the more than 100 "low-tech" people interviewed for the book, but I do want to acknowledge Kimberly Abbott, Christina Alvares, Ken Alvares, Leeba Aminoff, Lyn Anderson, Judy Barron, Sherry Borg, Donna Brewster, Bob Broenen, Lucy Caldwell, Matt Cavicke, Marla Cohen, Bob Daubenmire, Birgitta Delurgio, Amy Fowler, Christina Gage, Debbie Sue Hayden, Holly Hedeman, Jill Higgins, Stephan Keller, Trisha Keller, Jensine Kendall, Bernie Knost, Lyn Lasar, Barry Mainz, Roseanna Marchetti, Michelle McGarry, Dennis Mitzyck, Shannon Murray, Eliza Osborn, Kevin Roebuck, Garry Ronco, Aimee Ryan, Gretchen St. Lawrence, Allen Schloss, David Spenhoff, Brian Stahmer, Patrick Sullivan, Robert Todd, Cameron Truong, Mimi Tudor, Diana Wolf, and Wendy Yamaguma.

Very special thanks to Lucy Caldwell for being the world's most pleasant and energetic research assistant; it was a real pleasure to work with her.

Altair Literary Agency in New York is the home of Andrea Pedolsky, my agent, and her partner, Nicholas Smith. I've benefited from their deep knowledge of the publishing world, and I thank them for their support and their hard work on behalf of the book.

Wade Fox and Aaron Wehner, of Ten Speed Press, provided first-rate support and guidance in turning the manuscript into a book. I'm very grateful to both of them. I also thank Kit Kinrichs and Amy Chan from Pentagram and Lisa Patrizio from Ten Speed for bringing their top-notch talent to bear on the design of the book.

Scott McNealy is the president and CEO of Sun Microsystems, and therefore my ultimate boss. Thanks, Scooter, not only for being a superb leader (ahem!), but also for your comments on how nontechnical people get involved with and succeed in this industry.

Betsy Collard, of the Career Action Center in Cupertino, California, is an author and a true career professional. I've learned a lot from our meetings, and I'm also very grateful for her encouragement.

Special thanks to Duk Chun, Mark Opperman, Ted Tudor, and Andy Green. As Aristotle said, "Without friends, no one would choose to live, though he had all other goods."

Finally, special thanks and love to my lovely and patient wife, Gesine, who read every word of the manuscript and provided many corrections and thoughtful observations, from which the final work has benefited greatly.

INTRODUCTION

**Blessed is he who has found his work;
Let him ask no other blessedness.**

—*Thomas Carlyle*

This book was written because I spilled a glass of wine on my neighbor during a flight from Washington to San Francisco. Her bag was new but must have had wine-repellent properties because there was no sign of obvious permanent damage. Of course, I apologized profusely, which led to conversation, and, this still being North America, albeit at 40,000 feet, I asked her the standard question: "What do you do for a living?"

"I have a degree in art history," she said, naming a well-known Eastern college. "But there weren't any jobs in my field, so I started out substitute teaching for a couple of years. Then I got a job as a receptionist, and that's what I've done for the last eighteen months or so."

We talked some more. It was obvious that she was intelligent, well-educated, and articulate. "Do you have any long-term plans?" I asked. "You don't intend to work as a receptionist forever, do you?"

"I hope not," she laughed. "I'm not really sure. Back in junior year I knew I couldn't get a job if I stayed in my field, but I loved it so! Right now, I'm trying to save some money so I can get over to Europe this summer with my friends."

"Did you ever consider working in a high-tech field?" I asked. "The computer industry, for example."

"Are you kidding?" she said, her eyes widening. "I don't know anything about computers—no one would want me. I'm interested in something creative, and technology sounds so boring!" She smiled again. "I guess those are three pretty good reasons why I haven't."

A few months later I recalled this story when I met another young woman whose college major was in fine arts, and who was contentedly working in high tech. But at the time of the airplane conversation, I didn't have an example like this woman to cite to my co-passenger. In fact, I knew almost nothing about how people with no technical education got jobs in high tech—or even if there were many people like that in the industry. I knew that I was "low tech," with my degree in government and French literature from Columbia, but I thought I was an anomaly.

Still, the conversation stuck with me, so I raised the subject a few days later with a friend who was then a senior manager in a large computer company and who is now a VP in a young software company. I asked him what skills he considers fundamental to success in the high-tech industry. He's an engineer and very technical, so I was surprised at his reply:

> Excellence in communication. That means listening, writing, speaking—all aspects of communication. Listening is especially important because that implies a lot of social skills, being able to determine what a person is really saying beyond the words themselves, understanding what is driving him. Speaking skills are also important. What is the effect going to be on others if I say something this way or that way? Will it, for example, be seen as a challenge? What words do I choose? After all, that is what 95 percent of your people really do all day long.
>
> In many companies, in order to land the higher-paying jobs, people have to actively enjoy helping others do their jobs better, working through others by managing, coaching, and planning—all skills you would acquire from a classic education.

If communication is the single most important skill required by the high-tech industry—and by the way, over 60 percent of the people I interviewed spontaneously expressed this belief—it ought to stand to reason that, possessing this skill, anyone, including a fine arts major, ought to stand a fair chance of getting a job in high tech. Let's put it this way: If anyone who can communicate well can get started with a job in high tech—*any* job—he or she must have a pretty good chance of success. Sure, other skills will be needed, but the primary skill doesn't seem all that daunting.

The woman I had met on the plane had stated three reasons for not considering a high-tech job. First, she didn't know anything about computers. But how much does one really have to know about computers, at least to get in the front door and into a decent entry-level job? Aren't there ways to get started doing something in a high-tech company and then learn what you need to know after you are safely in your job? Is there some kind of short-term training you can undertake and then use to leverage yourself in? "They wouldn't want me," my co-passenger had said. Well, it's true that art history isn't in high demand in high tech—unless you're doing a CD-ROM on art treasures of the Louvre, for example—but that's not the point. It is a great mistake for art history majors to believe that the skills and knowledge acquired during study in that field are limited to that field. Exactly the same is true of all nontechnical majors—history, religion, urban planning, Slavic literature, anthropology, and so forth. My airplane companion was smart, articulate, and appeared to have a sense of humor—all highly desirable traits for workers in high tech. Would there be room for her creativity? Absolutely! The high-tech industry is changing so fast that there are almost no rules, and creativity, energy, and adaptability are in demand.

Oh yes, and high tech pays well enough that my airplane companion could retain art history as an avocation and actually visit the Pitti Palace, the Musée d'Orsay, even the graves of Sian.

Once I'd investigated the topic, I found myself wishing I could talk once again to my flight companion; I wanted to explain why looking into high-tech jobs makes excellent sense. First, many high-tech jobs require no technical knowledge at the entry level and sometimes even at fairly high levels. Second, the industry needs well-educated, creative people to fill these nontechnical positions because without them the technology won't make it out of the labs. Third, the high-tech industry is fast-moving, exciting, and fun. What would her reaction have been to my three points? I wondered. I suppose she might have had the same reaction you may be having as you read these lines—show me! Then she might have said, "Even assuming everything you say is true, what would it be like in such a place for someone like me? It'll be totally different. Who will help me get started?"

Pretty good questions. And as I thought more and more about it, I thought of more and more questions, many more than I had the answers to, and certainly more than could be covered in a flight between Washington and San Francisco. So I decided to write this book.

To get at some of the answers, I've interviewed well over a hundred "low-tech" people, and those interviews form the backbone of this book. I've heard some amazing stories, some of which I'm going to share with you. And every observation and bit of advice contained in this book either comes from my own experience as a "low-tech" person with some fourteen years experience in the high-tech industry or directly from the experience of others with backgrounds like mine . . . and yours.

This book can do several things for you. The first is to break down the misunderstanding and fear you might have that high tech is just not for you. I see this fear all the time, but there's an easy way out for many of us. Since we've already concluded (falsely) that to work in high tech you must be a genius—or at least very highly educated in the field of computer science—we don't really have to confront our fear of high tech. One method I use to combat this false notion, and to get you thinking positively and not fearfully, is to feed you hard information—lots of it—on what high-tech companies are like, what goes on in them, and what the jobs are, in order to demystify the industry.

Secondly, this book encourages—make that *compels*—you to get better acquainted with yourself and your strengths, and equips you with the tools to do so. After years of talking with job candidates and carefully reviewing their resumes, I continue to be amazed at how hard people are on themselves when it comes to evaluating their worth in the marketplace. It seems that such people are programmed from childhood to look at themselves supercritically and to denigrate their skills and accomplishments when they should be shouting them out for all to hear. The financial, familial, and social pressures that confront many of us as adults on a day-to-day basis conspire to reinforce this self-defeating behavior, and the result is a classic case of low self-esteem.

To secure that coveted high-tech job, it's essential that you become your own best salesman. That entails sloughing off all the pent-up qualms and doubts you may have been harboring for years, and redirecting that negative energy into identifying the hidden assets you can bring to the high-tech party. I have written this book with an eye to helping job seekers review their lives, abstract the relevant skills and experience, and present these in a compelling way—whether via resume, letter, phone conversation, or interview—that makes sense to a hiring manager and that matches the requirements for a particular position.

Chapter 1 provides some encouraging words for those of you who are sitting on the fence and considering a high-tech career. The demystification begins with chapter 2, which offers a brief introduction into what we mean by "high tech" and reveals where high-tech companies are to be found. It continues with chapter 3, "The Jobs," which starts with a listing of some sixty positions in high tech for which no formal technical training is required. You'll then listen in on some of the interviews I conducted with low-tech people who are working in many of these positions. They not only describe what the positions are like as seen from within, but also talk about how they found themselves in these jobs. I think you'll be pleasantly surprised to find that these are pretty normal people, a lot like you and me.

In chapter 4, "Starting Your High-Tech Career," I tell you how to launch, manage, and conclude a successful high-tech job hunt. As in the rest of this book, I've distilled and pass along to you everything I've learned over the past fifteen years about how low-tech people move into high-tech jobs and turn them into high-tech careers. The process of getting that first job is part science and part art. The science part comes in such areas as networking to develop those all-important connections, constructing a really good resume and cover letter, and studying your targeted potential employers before moving in on them. The artistic part comes in communicating with all the people with whom you will have contact from now until you settle back in your chair in your cubicle or office and put the first photo on your shelf—people such as temp agency managers, high-tech workers, recruiters, human resource personnel, and hiring managers.

To this end, chapter 4 emphasizes how important it is to map out a strategy and stick to it, making notes of your opportunities, contacts, what leads should be followed up, how your resume should be tuned for a particular position, and so forth. The chapter also contains useful information on how to develop a network in high tech, and the important role temping or contracting can play as a means of obtaining a permanent position. It also covers the role of headhunters, the Internet, and job fairs as part of the job search.

Optimistically entitled "Computers in Thirty Minutes," chapter 5 continues the theme of demystification. Actually, I suspect it might take you a bit longer to process, as it's packed with information. Pay special attention to the section on high-tech jargon at the end of the chapter. I'll let you in on a little-known secret. Being able to "talk the talk" is a key to being considered an insider. So, as amusing as some of the expressions are, to a degree they are like code words or secret handshakes. When you use insider jargon you're one of us. Another little-known secret: the main reason I want more liberal-arts people in the high-tech industry is that I need allies in my fight to save the English language. The words and phrases in the glossary in chapter 5 are in direct opposition to this goal.

Chapter 6 looks into the culture of high-tech companies. In one sense, generalizing about the culture is useful. Elements of the culture exist in virtually all high-tech companies. On the other hand, there are vast differences between the culture of a giant such as Hewlett-Packard and a small start-up company that's into Internet gaming. Before you take any job in high tech, you owe it to yourself to get as well acquainted with its culture as possible. No salary can make up for a bad fit between you and the company's culture. I promise you it won't work out.

Chapter 7 will help you understand how to survive and thrive in the fast-paced world of high tech. It offers advice on how to grow your job and how to turn a job into a career. Terrorized by stories about downsizing? Don't worry. Not only can humble low-tech people like us have careers in high tech, we are very much better qualified than the techies to keep our jobs when times turn rough. Low-tech people, once they are on the job, tend to be much more sensitive and responsive to the fast pace and twists and turns of the high-tech environment. They are more aware of the political elements that are present in any workplace, and many of them develop a keen perspective on which way technology is going. Low-tech people are harder to categorize than, for example, a programmer—so they find it easier to take advantage of training opportunities, spot career-enhancing opportunities, and move quickly to take advantage of them. You'll also learn about some of the psychological blocks and frustrations people like us have experienced in considering a move into high tech and what others have done to overcome these.

Chapter 8 looks at the realities of start-ups: how they get started, who founds them, when a nontechnical person is welcome, how to assess the chances of success and the consequences of failure. It was a fun chapter to write, since virtually everyone in high tech at some stage of his or her life has contemplated getting involved with a start-up. I think you'll find it interesting reading.

Many people want to work in high tech for a chance to go overseas in a well-paid position, with all the prestige and support derived from working for a high-tech company. How real are these expectations? What are the chances of this ever happening—and how can one make it happen? And what happens when it's time to go home? These are the questions considered in chapter 9.

Finally, in chapter 10, the good and the not-so-good aspects of high-tech work are summarized. I hope the discussion encourages you to consider high tech as a real opportunity open to most well-educated people who are flexible, tolerant, intrigued by the thought of working in a rapidly evolving and important industry, and want some excitement in their lives.

These chapters have enough solid information to encourage you to take a good look at the high-tech industry as offering an exciting and rewarding career path. Many others have taken a deep breath, then followed this path, and they have never looked back. You can follow in their footsteps.

CAN YOU HAVE A CAREER IN HIGH TECH?

I think it must be one of the best-kept secrets in the country: The high-tech industry, the economic engine that's been driving the American economy, really needs low-tech people like you and me. Oh, I know it doesn't look that way. If you open the paper and take a quick look at the classified ads, as I'm doing right now, you find descriptions for lots of job openings, but they don't seem to have much to do with our backgrounds. For example, here's a quarter-page ad that must have cost the company placing it a bundle. It lists ten categories of jobs, including software engineers with "strong C++ on Windows 95/NT platform, NT internals, COM/DCOM, TCP/IP, Winsock, multithreaded application development, database with ODBC." Another ad seeks "high speed digital communication engineers" to develop "implementation approaches and perform tradeoffs and evaluations of technologies . . . perform synthesis of DSP functions and algorithms into gate count, implementation technology and develop size, mass power estimates." And so on.

If you're hoping to make a career change, I can't think of anything more discouraging than turning to the help wanted pages. Yet, consider the educational backgrounds of a few friends of mine who are happily working in the high-tech industry in various places in the United States:

- Fine arts
- Art history
- Fashion design
- Urban planning
- English literature
- French literature
- History
- Theology
- Linguistics
- Music
- Psychology
- Economics

- Filmmaking
- High school diploma
- Philosophy
- Broadcast journalism
- Government

I asked a young woman whose undergraduate degree was in English, who is now the director of human resources for a fast-growing start-up, what, if any, skills she had acquired in her undergraduate studies that were directly relevant to high tech. She replied:

> I'm surprised to say there are more than I thought. One skill is writing. I end up doing a lot more writing than I ever thought I would have done. Obviously, I do offer letters; I do the job descriptions that get posted; I help managers write performance appraisals; and I've written biographies for our PR guy. I've written letters to our landlord, and I've written a lot of carefully crafted letters regarding employee issues. So the writing skills have been absolutely invaluable.
>
> And communication skills are another asset. We have a sort of joke here. If you're dealing with engineers, you have to grade on a curve for communications skills. Communication is less significant for the technical fields than the nontechnical. My theory is that, as an engineer, your work results are very tangible—they are either good or they're not. In other fields they're much less tangible, so the [communications] skills you have are much more important.

A young man who is in channels management at a very large computer company observed:

> I started the job knowing very little about their products, other than knowing how to use a computer. I didn't understand anything about their distribution strategy or much about their product line. I was clueless, actually. I learned on the job, just picking it up on the fly, making it a point of doing some research every day. I made friends with the product people so I could understand when a product was coming out and what their strategies were.

Another liberal arts graduate who has been in the industry for six years said:

> I'm fascinated with the high-tech industry. And if you find this industry fascinating . . . that's enough to be in this industry and be successful.

And a woman whose Spanish language skills took her from work as a homemaker to an international high-tech job advised:

> Definitely don't be afraid of high tech. It seems to be very daunting at first, when you don't know anything about it. . . . But if you can use a telephone, you can use a computer. If you can use a tape deck, you can use a computer. I don't think people should categorize themselves as high tech or low tech. We live with technology all around us, and we're all versed in it to some degree.

What kinds of jobs do such people manage to sneak into? How do they pull it off? Are there really career paths for people like them? What happens to them when things turn sour and there's a "downsizing"? Are they really doing important stuff, or are they just peripheral to the main activities of the high-tech company?

If there's one central theme to this book, it is that high tech offers career choices to a very wide spectrum of people, most of whom probably can't imagine that they could qualify for a position in the industry. I don't want to gild the lily too much. To get a job in high tech requires determination, organization, and persistence. If you've been in a different field for many years, the gap you'll have to bridge may be wide, but it can be overcome.

Sometimes everything just seems to fall into place, as it did for Jenny. She's about thirty-five years old, and until 1995 she was a waitress in restaurants in Santa Cruz, California. Here's her story, in her own words:

> I have a bachelor of arts in finance and a minor in German. I started waiting on tables after graduation [in 1985] and just kept it up. I waited on tables, and I managed a gas station. At the last

job I took a table for a friend of mine, who just did not want to take this last table. I struck up a conversation with this man, and he asked me for a resume. I said, "Yeah, sure," and of course I didn't do anything about it. Then a week later he came back to the restaurant. "Why didn't you send me your resume?" he asked. So I did.

This man was with a tiny start-up, and the start-up was looking for people who did HTML for Web pages. HTML was still new at that point.

The day before the interview I got a book called *HTML for Dummies* and I stayed up all night, the night before I went in. I think the reason he asked me to apply was our reasonably intelligent conversation at the table, and that I tried doing even my menial waitressing job well. I worked hard at what I did.

I don't suggest that you drop what you're doing and rush off to apply for jobs in restaurants on the California coast. Waiting tables is not a sure ticket to high tech, but at the least, this story illustrates that opportunity can appear in many different forms. Jenny changed companies after a year and went with one of the Silicon Valley giants in a job that has her developing corporate presentations and working on the company's huge Web site. I asked her how the money compared to what she made as a waitress. Her reply:

I was doing pretty good waiting tables, but I'm doing a lot better now, making about three times as much.

Arthur has a somewhat different story. He's in his mid-forties and has been in high tech for about a year.

I felt like a fraud at some time or other in half the jobs I've done. I was a teacher for grade school kids. I taught dance lessons. I did theater technical work. The main thing I did was produce children's videos for about five years. I had studied biology and chemistry, and then I jumped around, did photography and all those other things. I found that I can land jobs for which I often don't feel qualified. When I got to [the high-tech company] I realized that nobody knows everything about this stuff. Almost everyone out there is in a similar state. It's like just getting over that feeling that you don't know, being able to deal with that and understanding that you don't need to know everything.

[My decision to try high tech] was motivated by a couple of things. It was financial, and I also felt like I'd missed something. A lot of fellow students went off in the mid-eighties and worked for Apple and some of the other high-tech companies. I thought, oh, well, they're doing corporate videos, and I'm doing educational stuff. Then the educational market died, and I thought, now I'm too old. I felt like I didn't have connections any more.

Then I was hired for a contracting job through someone I knew at Apple. I had an understanding of film and video, but I didn't understand networking very well. I guess it was just my life skills that got me through the job, my ability to talk well to people, to listen to people, and to be present. I did this contracting work less than a year.

Then I received a call from someone I'd met at Apple, and he invited me to apply at [present company]. There was plenty of work and opportunity for growth. I had done a lot of work at the university where everything was so traditional, so limited, with so little money or opportunity. And here was a place where I felt almost anything could happen.

CAREER CHOICES

This book is about career choices. It's written for people who want to make a career change and college students or new graduates who are wondering what to do with their lives. Some of you may have been in the workforce for only a couple of years, others for much longer, and still others not at all. However long you've been at it, if you are typical of the many people I've counseled over the last few years, you've started to question the choice or choices you've made up to now. In fact, if you've really given the matter some thought, you may be wondering whether life ever really gave you any choices. You have plenty of company. After working for a while, many people start to have doubts and qualms. They start asking themselves how they ended up where they are and whether their daily routine is all they can expect to get out of a working life that can last thirty to forty years or more.

Concerns about money start people thinking, but often that's not it.

Instead, it's that elusive thing called *job satisfaction*. Most people I've met would gladly sacrifice some money (not a lot!) if they could only get a job they'd enjoy. How do we get into this sort of rut? Often, we've taken a career path that others "expected" of us. Or we had to start paying back those school loans somehow, so we blindly grabbed the first job that came along. Very often a concern for security (often foisted on us by others) dictated the field we chose to work in. Idealism, too, may have played an important role—the desire to give something back to society or somehow make a difference in people's lives, rather than just grubbing for money.

Why do people stick with jobs they don't like? Why aren't they even aware of the choices that are out there? These questions have bothered me for years because I see a lot of unhappiness that I happen to think is completely unnecessary.

THE HIGH-TECH INDUSTRY AS A CHOICE

In my experience, the hardest leap of faith for nontechnical people to make, be they recent college grads or people who have been in the workforce for a few years, is that they really are needed in the high-tech industry. So let's first address this concern.

High-tech companies have some really smart people who design their products, which for the moment we'll classify as either hardware or software. People who design hardware and software are the sorts of people high-tech companies advertise for. You're forgiven if you are one of the many hundreds of thousands of people who've been convinced by such ads that there's no place for them in high tech. Well, there's something you should know. Newspaper ads cost a lot of money. So high-tech companies will only advertise for the positions that are the hardest to fill. Feel bad that you don't qualify for such esoteric positions? Shake that negative feeling off right now. Few companies advertise for the nontechnical positions, but they are there. The truth is that good people are hard to find, even in times of economic downturn. The practice of not advertising for such positions plays squarely into the hands of people like us, who make a plan for getting into high tech and then execute that plan faithfully.

High-tech companies also resort to headhunters and recruiters when they want to find someone for a high-level job. For example, a software company might be looking for a very experienced director of product marketing. During the downsizing craze a few years ago, human resource departments were cut to the bone, and since then they've been kept pretty lean. So companies rely on outsiders to find upper level managers; the companies just don't have the staff to advertise for and evaluate lots and lots of applicants. And you, looking for a position in high tech, usually can't exercise any influence on such outsiders: They either call you, or they don't.

But if you're after a job from the entry level to just below the top (say, a manager's position just below the director level), you can control a lot as you start your move toward a career in high tech. You'll learn about the job possibilities in chapter 3 and how to get a job in chapter 4.

The jobs you might fill are ones technical people can't do and are often just as important as the technical jobs. The designers and developers think up new technologies and products, but that's all they do. They do not examine the marketplace to determine what the needs are for such products, and a company that keeps making things for which there is no need is obviously doomed to fail. So how do companies determine what the marketplace needs? Well, they send people like you and me to go talk to other people, people in banking, for instance, or in education or transportation, entertainment, or oil drilling, or . . . you get the idea. Sometimes technical people are given such assignments, and often such an arrangement doesn't work out. Why? Because what is called for, above all, is skill in communication. And as wonderful as technical people are, they generally aren't blessed with superb communication skills.

If you've reached the stage where you're thinking seriously about a career in high tech, you may find yourself facing some rather powerful psychological hurdles. You'll have to overcome these, and they can be as strong or stronger than the external hurdles. The first hurdle may be a sort of natural inertia that keeps steering you back into the same old rut you've been following for years. Your present work may not be satisfying, but it's what you know. It makes few demands on you, except stamina and a tolerance for boredom. Your job may not pay all that well, but you feel it offers security. It may have little status, but it contributes toward making the world a better place, perhaps by helping less fortunate people. And you may have been contributing for years to a retirement plan that would be at risk should you take the plunge and really swing into action to change careers.

Making a career move can be risky, and many people are quite risk-averse. What if you make the wrong move? What if you wake up a few months or years hence to the realization that the whole thing was an awful mistake—one that it is now too late to rectify?

Of course, you, and only you, are responsible for your life and what you choose to do with it. But to introduce a little perspective into the question of risk, I'd like to point out that job security in every field is far from certain. Teachers get fired. Actuaries get fired. So do salespeople, financial analysts, clerks, telephone linemen, and civil servants. Perhaps all of us should start thinking about alternatives to what we're doing today.

A man I interviewed for this book, who joined high tech at the age of forty-five, said to me:

> When I look back, I see so clearly how I gave myself negative statements and how I held myself back. Yet I was still somehow able to get through that and find myself in the high-tech field . . . where there is a tremendous amount of opportunity, where you can be rewarded for your tangible skills and your abstract abilities, and rapidly. You don't have to stay stuck. I don't think high tech is for everyone, but my experience with it after just under a year is that it's tremendously exciting.

I have a friend who used to work for a large East Coast architectural firm. He isn't an architect, but was involved in the marketing and sales side, as a writer of many successful proposals for large architectural and urban planning and design projects. He lost his job a few years ago when the economy went downhill, but the firm kept him on as a contractor at pretty much the same pay rate, only without any benefits.

After a year or so, the contract work had dwindled, and so had his income. He was able to get the occasional bit of work with other firms, but soon this, too, dried up. When he could no longer make the payments on his house, he woke up. Too bad he hadn't correctly assessed his situation and devoted some time to planning his next career. He ended up managing a health club.

Another phenomenon that starts people thinking about career alternatives is burnout. This condition can afflict persons in almost any field, from police work to psychology, from nursing to teaching, from investment banking to social work. Who doesn't know someone for whom the daily routine has become so oppressive that he or she is ready for a drastic change in work?

A great motivator for making a career change can be the prospect of making more money—perhaps a lot more money. Social workers, for example, perform very important functions in our society, and society should honor and reward them accordingly. But it doesn't. It expects them to soldier on at $18,000 to $20,000 per year. At that rate, though, they're often better off than those we entrust with caring for our own children while we, the parents, are attending to our own lives. The turnover in children's day-care centers is over 40 percent each year, which is not surprising considering that salaries in that area are among the lowest tenth for all wage earners. Even when society accords respect to a profession, financial rewards don't necessarily follow. Just look at teachers' or university professors' salaries. Of course, one must really believe oneself worthy of making more money. A lack of self-confidence often keeps people from seriously exploring better-paid career alternatives.

I'm not suggesting that all underpaid social workers and college professors throw over their careers and try to move into the high-tech industry. All I'm suggesting is that they, and you, have choices. I'm suggesting that if you feel burned out or underpaid or undervalued or simply ready for a change, you haul yourself out of your rut and seriously consider making that change.

About a year ago I was on a flight from London to San Francisco. Sitting next to me was a businessman, an American. We introduced ourselves and started to chat. It turned out that we both had sons, and after a while he shared a deep concern with me. His son, whom we'll call Ben, had graduated from a small and not particularly distinguished college. Ben's love was jazz; he played several instruments very well and had organized a band that traveled around California playing gigs. That was what he'd done since graduation. It was a large band, too—thirteen players.

The problem was that Ben was now twenty-nine years old, and he had never cleared more than $7,000 or $8,000 a year with his band. The only way he was able to survive was sharing a tiny room with his girlfriend, who also made only a few thousand dollars a year. Ben wanted to make a change, but he couldn't figure out what to do or who would want him. His father said, "I wish I could think of something for him to do."

I'd already given the father my card, so I said, "Well, if he's interested in looking into high tech, tell him to give me a call. The least he'll get out of it will be a free lunch."

To my surprise, Ben did call me a couple of days later. This was a good sign because the motivation to make a career change is the most important element in getting the process underway. When he showed up for the lunch meeting, he had (at my request) brought a resume he'd written some months before. It started out by giving his name and address at the top and then listing every position he'd had in his attempt to survive—positions such as busboy, waiter, and bartender. Every beginning and ending date of employment had been carefully noted. Way down at the bottom of the resume, under "Interests," was his band.

As I listened to Ben, prompting him with questions, I realized that this young man had a number of notable accomplishments. First, organizing the band was his idea. Second, he'd recruited all the musicians. Third, and even more important, he'd managed to keep them all together for several years. Next, he'd organized all the band's engagements, collected the money, booked rooms for the tours, paid all the expenses, did the public relations, and paid everybody their share of the revenues. In short, Ben was a manager. He just didn't know it.

I wasn't sure how successful Ben would be, and frankly, I thought it was a shame that he wanted to get a "real job" because I knew the band could not survive his departure. But he wanted to try a career somewhere, so we worked on his resume. We rearranged it to feature the creation and management of the band, in terms that I thought a hiring manager would relate to. As for the bartending jobs, we disposed of them in a single sentence, to put them into the proper minimal perspective. The strategy was to get him into high tech via a temp agency. I told him how to set up an appointment and asked him to be sure to let me know what happened.

A week later I got a note from Ben's mother (the parents were divorced). She wanted to thank me for working with Ben, who had landed a one-week temp job in a software company. It wasn't the new job that made her so happy; it was that someone had helped her son discover some—certainly not all—of his skills and abilities. A week after that I got a call from Ben. He had been shifted from his initial one-week assignment to another company, where he was engaged for three weeks. He wasn't sure where all this was going to lead—after all, he was only doing somewhat menial kinds of work—but he was making more money than he had ever imagined possible. Furthermore, his co-workers were great guys, and he was intrigued by the environment.

A couple of weeks after that, Ben's father called. He told me that the manager in the first company, where Ben had only worked for a week, had called the temp agency and had told them that he had to get Ben back, no matter what. It wasn't just that Ben had done a great job during the week he was there. The manager happened to be a jazz fan and missed talking about it with Ben. Sounds like a great start to a new career.

You never know what may be waiting out there for you, so I encourage you to start exploring for yourself. The president of Sun Microsystems, Scott McNealy, who incidentally does not have a technical background, describes working in high tech in these terms:

> You've got to have an intellectual curiosity. You've got to have a desire to make more than just money. You've got to want to improve the standard of living, the tools that we have, because that's what technology does. You're going to need some psychic income because it's hard work, but it's exciting. Sometimes I refer to it as too much chocolate cake.

Ready for some chocolate cake?

WHAT IS "HIGH TECH"?

The term *high tech* is used these days to describe lots of different things. For example, if you're lucky, your dentist will have an ultra-high-tech installation for working on your teeth. The benefits to you are better treatment, better teeth, and a happier, painless experience every time you get into that contoured chair. High-end toasters are high-tech, too. They come with tiny chips in them that regulate their function and that can distinguish between a slice of rye, a slice of Wonder Bread, and a genuine bagel (a disappearing breed). Moving up the complexity ladder, all kinds of medical devices are high tech as well, from heart monitors to CAT scanners. Cars, too, are high tech. Gas consumption, braking, air-conditioning are obvious examples of systems involving high-tech components, but a friend of mine, who knows about such things, tells me that certain luxury makes will soon have up to 150 microelectronic chips in them.

Homes, which at the beginning of human history were just caves with a few animal skins to dress up the premises, have become dens of high technology. The toasters are there, of course, but so are the security systems, the multimedia equipment, the PC/fax/copier/scanner, the Web phone, and the black box that turns your TV into a gateway to the Internet. The outdoor spa, or hot tub if you prefer, which to the Woodstock generation (if you don't know what this refers to, please ask your parents) was literally a simple wooden tub into which you first put hot water and then yourself, today has the electronic complexity of a small airplane: sensors, timers, temperature regulators, digital displays, lights—you name it. An article in *Business Week* states that "several years from now, the computing landscape could well be dominated by smart devices, new kinds of networks and embedded systems—in cars, buildings, interactive television sets, and credit cards." That's a true statement, except it won't take several years for it to happen. In a fascinating article in the August 3, 1998, *Fortune* entitled "The Network in Your House," Erick Schonfeld looked into the way computer networks are penetrating into people's homes, serving both as a gateway into the Internet and as a means of serving as "command centers" for regulating every device in and around the house.

When this book uses the term *high tech*, it refers to the industry comprised of a very broad group of companies. These companies range from the makers of desktop computers—PCs—for home and industry, such as Dell, Apple, and Hewlett-Packard, to manufacturers of powerful computers called servers, such as Sun Microsystems and IBM. On the software side, the spectrum runs from giants Microsoft, Sun, IBM, Compuware, SAP, Oracle, and others, to smaller companies such as Intuit, or individuals who develop applications for specific niche markets. There are about 5,000 Internet service providers, or ISPs, in the United States today, ranging from America Online and Compuserve, who provide gateway access to millions of subscribers, to mom-and-pop businesses with a few hundred subscribers. There are the companies that make the "search engines" that hunt for specific pieces of information on the Internet, like Yahoo! and Excite.

Then there are all the companies that provide peripheral equipment, such as modems, scanners, printers, fax, and networking cards for PCs and laptops, and game-related software and hardware. A whole branch of industry is concerned with networking hardware and software—things called bridges, routers, and switches—with Cisco Corporation being the current leader.

There is a publications industry that serves all these companies, as well as consumers, or "end users" in high-tech lingo. These include a myriad of magazines and technical journals, many of which are available at your local bookstore. There are market research firms and PR firms who specialize in high tech.

Through a process known as "convergence," the telecommunications, retail, entertainment, and banking industries are being drawn into the world of high tech. Many excellent careers will come about as people with liberal arts backgrounds and some years of experience in today's high-tech environment leave the high-tech industry, using their experience to take high-paying jobs in those fields. Such people have an advantage over the purely technical people because they do not have "programmer" stamped on their foreheads.

High-tech companies are located in lots of places in the United States. They run the gamut from one- or two-person software development firms to multidivisional giants with tens of thousands of employees and sales in the billions of dollars. Of course, high-tech companies are found in many other countries as well, but if there is one area of business in which the United States is clearly the world leader, it is the high-tech industry.

WHERE ARE THE OPPORTUNITIES?

Where are high-tech companies to be found? A few years ago this question was a no-brainer. The answer was: in Silicon Valley, south of San Francisco, and in several communities to the west and north of Boston, Massachusetts—oh, and a few in the area known as Research Triangle, near Raleigh-Durham, North Carolina.

But that's no longer the case. High-tech companies, particularly software companies, can be found in the Northwest, in Texas, Utah, Colorado, New Jersey, New Hampshire, and many other states. You can rather easily locate them by consulting any standard commercial directory, such as *Ward's Business Directory* or *Rich's High-Tech Business Guide*. There's a bit of a caveat, however. The job possibilities for low-tech people may be very limited in areas that don't have corporate headquarters. Sales offices, for example, are located everywhere. But you don't stand a chance of getting hired into an entry-level sales position in such offices, and they don't offer any other kinds of job except for administrator positions. Your best chances are in an area that has lots of corporate headquarters and therefore offers a wide variety of positions and levels.

THE ORGANIZATION OF A HIGH-TECH COMPANY

High-tech companies, regardless of how their organization charts may differ, have a structure that supports three distinct types of activities. First, there is the research and development area. This is the real techie stuff—

the actual design of hardware, software, networks, whatever. The second area involves all the myriad of activities needed to shepherd the technologies or products out of the labs where they were designed, getting them into the marketplace and, hopefully, into the hands of customers. This area has many career opportunities for nontechnical people. The third area consists of those ancillary or support functions that are required to keep the other two going: legal, finance, human resources, investor relations, public relations, facilities, and so forth. This area also holds many opportunities for low-tech people. Chapter 3 presents a fairly comprehensive listing of the positions available in the latter two areas, and in many cases, I've described what a particular position actually entails. As mentioned previously, where possible I've also included observations by low-tech people currently working in such positions. This is the best way I can think of to give you the true flavor of what it's like to work in high tech.

Within the structure of the corporation, high-tech companies have a pecking order. Each function shown on a company organization chart will be staffed with people at different levels. For example, a basic position in marketing communications, or "marcom," might be a "level two" in the particular job classification system of the company, whereas a group manager position might be a "level twelve." But the fact that there is a pecking order in a bureaucratic sense doesn't necessarily imply that a lot of pecking goes on. Everyone is too busy to devote much time to that sort of stuff.

WHAT'S DIFFERENT ABOUT THE HIGH-TECH COMPANY?

Chapter 6 gets into more detail about the culture of high-tech companies, but the high-tech industry has some fundamental differences it's important to know from the start. I think the major difference between a computer or software company and any other kind of company is the pervasiveness of awareness of technology. No matter what your job involves, you are constantly aware of the technology of your own company and where it fits in the total high-tech world. Finance people in a disk drive company don't just balance the books; they can talk about their company's products. They can also talk about domestic and foreign competition, and if you are a prospective customer and sign a nondisclosure agreement, they can probably give you a pretty good high-level briefing on future products as well. Contract administration people in a software company can do the same. Everyone has an opinion on where the industry is going. As an employee, no matter what your job may be, you find that you read stuff about technology in magazines, newspapers, and on the Web that you never thought would be of the slightest interest to you. Your significant other may try to restrain you, but you just can't stop holding forth at parties on the state of the computer wars or of database technology or C++ versus the Java programming language or whatever. I don't think that goes on in the toothpaste industry or the insurance industry.

Let's face it—high tech is sexy. We may bore the pants off unwary guests who thought they were coming over for a few drinks and a video, but at a minimum they stagger off down the drive muttering, "Jeez, I didn't know she knew so much about that stuff." And at a maximum, they may ask you to advise them on high-tech investments (don't give it). Everyone knows how important high tech is to the American economy. And if they don't, you'll soon let them know. For example, industry software revenues grew from $39 billion in 1990 to $122 billion in 1997. That's a pretty good growth rate, and it's increasing.

I really liked what a young Internet software salesman told me when I asked him what he liked most about his job. I expected, frankly, for him to talk about money, since filthy lucre has always been a prime motivator for sales folks. Instead he said:

> What I like about this job, this company, is that we're still defining the market. We're not entering into a set of standards that have existed for decades or longer. We're an industry that by definition is changing all the time. This fact enables us, as a company . . . to set the direction of the marketplace. We're constantly talking to customers who have new ideas about how to market themselves or a concept or a product or even their business ideas on the Web, and we're able to consult with them about how our technology would help them.

Another thing refreshingly different about high-tech companies is a sort of egalitarianism. It is a striking feature of the industry. High tech is pretty close to being a meritocracy. Yes, there are glass ceilings here and there, and as with every human endeavor, you find some pettiness, some prejudice, some stupidity—but not very much. What we are doing in high tech is, once again, too fast-moving, too demanding, and too much fun for us to get hung up on irrelevant social, racial, and class distinctions—who's eligible to play golf at such-and-such a club, for example. I've never worked in the automotive industry, but I understand it's big on executive bathrooms and exclusive golf clubs. I've yet to see an executive bathroom in a high-tech company, but of course they may just be well hidden, like special airline lounges in airports. Maybe the automotive execs have missed the wave because they were hanging out in their nice bathrooms. As for the golf clubs, I'm not a golfer, but in the Silicon Valley area, if you've got the dough, you can join any club you want. A friend of mine sells software, successfully, and his idea of a nice golfing weekend is to fly to Scotland and play at St. Andrews.

Everyone in high tech is on a first-name basis with everyone else, at least in the companies that count. That, too, contributes to a feeling of egalitarianism. And there is a genuine feeling of respect for what one's co-workers, be they above or below you on the org chart, do for the industry. There's something called a "spot bonus," which is money that can be awarded to someone on the spot for a bit of outstanding work. All it takes is a manager's OK, and it's done. It's great to be able to say thanks to a top performer, and emphasize the point with a $200 gift certificate to Nordstrom or $150 for a dinner for two at a local restaurant or a check for $500.

Even without this kind of largesse, there's respect. I have never heard anyone speak disparagingly of any other position in high tech because each of us knows there's a reason for that position, and that it contributes something to the overall success of the company. No one ever says, "Oh, she's just in marcom," or, "The ISV relations department is a waste of time." (We might agree that an individual worker is a turkey, of course. The right to think and say that is in the Bill of Rights.)

3

THE JOBS

What are the kinds of jobs that you, as a low-tech person, can reasonably hope to fill, both when you first enter the industry and later on, as you develop your career? I asked an experienced management trainer at a large computer company what jobs could be handled by people with a non-technical background. Her comments are encouraging:

> Actually it's quite open. As long as you have access to experts, meaning the folks that are technical, and you know how to communicate with them and you know whom to contact, you should be able to do most jobs.

The only jobs that are absolutely closed to people with a nontechnical background are the R&D positions; unless you are a self-taught genius, you simply won't be able to design integrated circuits or complex software without having put in a few years of formal study. Some of the professional positions are also closed, unless you have the right ticket. You can't get on the corporate legal staff, for example, unless you can wave your law school degree around. Similarly, some finance jobs will require that you be a CPA or that you have an MBA and a command of international tax questions.

But everything else is possible. Sure, you won't go sailing into many positions without gearing up for them. You have to go about it in the right way, and with patience. And that's a good thing, because getting there is more than half the fun; it involves a whole process of self-discovery and development, plus the joyful amazement when you realize that you've moved into a totally new work life and that you like it and it likes you. We'll look at what this voyage involves in the next chapter. Right now, let's look at the jobs. In many cases I've tried to go beyond the title and formal job description because I feel that often these don't have all that much to do with reality. I also thought it would be useful, and I hope enjoyable, for you to hear what low-tech people who are actually in some of these jobs think of them.

When I started thinking about this jobs topic, I thought that presenting it to you would be a breeze. Surely all jobs in the high-tech industry could rather easily be divided into "hard-tech" and "soft" groups. The hard-tech jobs would have to do with the guts of the hardware and software, whether designing, building, or repairing them. The soft jobs would be those being done by people who didn't have technical degrees. As I started to interview lots of people in the industry, I realized that things were not so straightforward.

First, I was amazed at how the "soft" jobs in high tech are related to the technology and its evolution. People occupying such positions often describe themselves in a somewhat self-deprecating way as "nontechnical," and perhaps it is true that they have had no formal technical training, but many, even most, of the people I interviewed for this book had learned a lot about the technical aspects of the company's products. In fact, they had to do so in order to do their jobs well. Take the job of public relations. If your image of people in these jobs is of flacks who are hired to grind out whatever material will induce people to buy a product, you haven't met a PR person working in the high-tech industry. He or she works on a team in which everyone is expected to get up to speed on the products: what the products do, their "positioning" with respect to the needs out in the marketplace, and what the competition is doing.

The second thing that surprised me was how many people there were in admittedly technical positions who came from nontechnical backgrounds—some people without a college degree at all. Perhaps that ought not to have been such a surprise. After all, this is the industry of Bill Gates, who dropped out of college to build a software empire out of Microsoft and along the way became one of the richest men in the world. But I wasn't prepared for the young man who was the product manager in a large company for a very technical product, whose college education had been almost entirely in the field of early church history, or the man who majored in Sanskrit studies (I'm not making this up!)

who also ended up in product management. One woman who had graduated in fashion design and marketing but tired of dealing with fabrics enrolled in a hardware sales training program a few years ago and now pulls down about $130,000 a year helping an international banking organization make sense of its computing systems. A former small retailer, with no college degree but a considerable talent in graphic art, now deals every day with software development engineers and creates successful ad campaigns.

So here was a paradox! The lack of a technical education didn't prevent these people and many others from getting jobs in the high-tech industry, yet when you really took a look at what they were doing, they had achieved a considerable (and to my eyes, an overwhelming) command of the technology. For some, it was necessary; but others chose to do so as a way of "growing" their jobs and making themselves more valuable. And in a rather heartening number of cases, they learned about technology because they just became fascinated with it and found it was pretty easy. One thing became apparent. One can't blithely classify jobs as technical or nontechnical and simply encourage and coach people on how to apply for the nontechnical jobs. To a degree, all jobs in high tech have some technical aspects.

When we talk about technological aspects of traditionally nontechnological jobs, we're talking about "high-level" technology. Chances are that, unless you're a genius, without a degree in technology you'll never design an ASIC, or "application-specific integrated circuit," but you will be able to know what the term means and use it intelligently in discussions with engineers, salespeople, or whomever. You may not understand precisely how multitasking operates, but if you know that it keeps your computer screen from "freezing up" so that you can't move on to another operation until the earlier process has finished, you're on your way to achieving a high-level understanding of the technology. Along the way, you will gain tremendous respect for the women and men who dream up and develop this stuff, and the wonderful, mind-blowing thing is that you will gain new respect for yourself (and respect from your friends) when one day, not too long after starting your high-tech job, you start to understand the technology. Achieving high-level knowledge of technology is not only fairly easy, it's actually fun. As a friend of mine put it, "My God! I think I need professional help. This stuff is starting to make sense to me!"

What do all these jobs pay? It's not possible to be terribly specific, because there are so many variables, but to give you an idea, on April 21, 1998, the *San Jose Mercury News,* quoting a survey conducted by the American Electronics Association, gave a range from $39,500 to $121,200 a year for Internet- and Web-related jobs. In Silicon Valley, project coordinator positions can range from $30,000 to $60,000 per year, depending on the level of the job, the range of responsibilities, and the experience required to do the project. Marketing positions run the gamut, from $35,000 to $150,000 per year.

Salary standards and the cost of living vary from one part of the country to another. Each position in the high-tech industry has a salary range, often with a huge difference between the lowest pay and the highest for any given job. Also, companies have different compensation philosophies; some make it a point to pay at the top end, while others are content to rest in the top quarter or top half of what the industry is paying. Obviously some companies pay in the bottom 25 percent. Unless such a company is a start-up and is willing to give you significant stock options, it's probably not a good idea to start out there.

Suffice it to say that the high-tech industry in general pays very well, better for a given level of experience than any other I know of. And there are usually unparalleled opportunities to continue one's education at company expense, both in formal degree programs and by taking courses within the company, which I count as nontaxable income. And, finally, the mobility within the industry, where a receptionist can become a marketing manager and a PR writer can move into sales, is another untaxed benefit. I'm not sure how or whether this can be quantified, but I suspect it's worth a lot to most high-tech employees.

A LISTING OF "LOW-TECH" POSITIONS

Here is a list of activities that take place in high tech. In some cases I've given the title of a position rather than the name of an activity. You'll notice that I don't follow this list with a description of every single job. Instead, I've concentrated on the major jobs for which no formal technical education is required. I wanted to give you the full list, however, because in your job hunt you will certainly come across many of them, and you should not hesitate to investigate further. The list isn't definitive because different companies use different names for the same activity, and sometimes the same name for very different activities. For example, business development in some companies involves acquisition of technology, or buying whole companies. In others it means a sort of sales job. And in my company, Sun Microsystems, it means both. An asterisk before a job indicates that it is covered in-depth in the following pages.

*Account management
*Administrator
*Business development
    (market development)
 Channels communication
*Channels management/
    development
 Channels recruitment
*Channels sales training
 Channels technical training
 Company education and training
*Contract administration
*Contract negotiations
*Corporate communications
*Customer education and training
*Customer response center
*Customer service business manager
 Customer service escalation
    manager
*Customer service representative
 Finance
  AR/AP
  Controller
  Credit manager
  Payroll
  Shipping/receiving
*Forecasting (orders and shipments)
*Human resources
  Benefits

 Generalist
 Investor relations
*ISV relations
*Marketing
  Advertising
  Brand management
  Catalog design and publication
  Competitive analysis
  Demand creation
  Event management
  Field marketing
  Field sales communication
  Marketing collateral
  Price book management
  Pricing strategy
  Product marketing
  Product packaging
  Product queries
  Sales kit preparation
  Media specialist
*Operations
  Compensation management
  Specialist
 Partnership manager
 President
*Press queries
*Press releases
 Professional services business
    manager
*Project coordinator

*Public relations
Publishing *Sales representative
*Sales support
*Strategic alliance manager
*Technical editing and publishing
*Technical writing

*Telesales
*Training
Webmaster
Web page art director
*Web page designer
Web page manager

When you are networking to find a job in the high-tech industry, which you'll learn how to do in the next chapter, you should see if the positions you hear about match, or sound like, the ones on this list. But what do these jobs consist of, really?

Account Management

Account management, sometimes called "large," "major," or "national account management," is usually a responsibility given to someone who has been working for a few years in the company. Generally speaking, account management does not require deep technical knowledge. This function really involves giving the most important customers a single point of contact within the company for the resolution of business issues. What is most important is that the customer feels that someone in the high-tech company is dedicated to making the relationship work well and is always available by phone, e-mail, or fax. Being responsive is an important element of being successful as an account manager.

This function also gives the management of the high-tech company an unbiased view of how relations between the company and its customer are evolving. Because the account manager's salary is not usually geared to quarterly sales numbers, his or her recommendations to management regarding all aspects of the relationship are more likely to reflect a concern for the long-term company goals. Sales representatives may actually have more frequent contact with the customer, but this contact is aimed almost exclusively at achieving quarterly sales goals. A sales rep who is having trouble making his or her numbers, may devise schemes for accomplishing this that serve the sales rep personally quite well in the short term, but may adversely affect the company-customer relationship over a period of time—for example, selling a large quantity of a certain item that the sales rep has reason to believe may be going "EOL" (end of life) in a few weeks.

Since high-tech customers do not like to get stuck with obsolete goods and usually insist on returning them anyway, in such a case the account manager might work with the customer and the sales rep and his boss to arrange an easy transition from the old product to the new, perhaps planning for a low- or no-cost "upgrade."

The account manager usually has contact with a higher level of management than does the sales rep, though this is not always the case. Account management is considered a headquarters function, but in fact, the people who carry it out are sometimes located at the regional sales office closest to the customer's corporate headquarters. Quite a bit of travel is involved. Large accounts may have several locations throughout the country, or even overseas, where various important issues may arise. The account manager must be available to help resolve these issues.

Account managers are sometimes recruited from the industry they will deal with as an employee of the high-tech company. Someone from the oil and gas industry, who knows about its data-processing needs, may end up working for a high-tech company serving as account manager to one or more oil companies. But sometimes assignments are made on a geographical basis: an account manager may take responsibility for three or four major accounts in the Pacific Northwest, for example, and these accounts may be in quite unrelated industries.

Administrator

This position is of considerable interest to nontechnical people, since getting hired into an administrator job is relatively easy. On the surface the administrator may seem to do for high-tech companies what secretaries do for other companies. In fact, the positions can be quite different. The traditional secretary has a boss; the administrator is a member of (and "supports") a team or workgroup. Some "admins"—the short term by which those occupying the position are almost exclusively known—complain that they have not one but several "bosses."

The high-tech company doesn't have some of the tasks that secretaries carry out in other industries. For example, managers in high-tech almost never write letters. Communication is carried on by phone, e-mail, or fax, all of which can be done directly from a manager's or individual contributor's desk. When letters are required they are almost always written by the sender on a computer and printed out directly or sent to the admin for printing out on company stationery. Appointments can be made by computer over the network, and since high-tech managers and others can consult each other's appointment calendars in this way, the admin's intervention is often unnecessary. But some routine tasks almost always account for a large part of the admin's working day. These vary but may include arranging customer visits, setting up large meetings, arranging travel for members of the workgroup, preparing overheads for presentations, delivering interoffice mail, working at trade shows, ordering supplies, setting up lunches, and so forth.

In the best high-tech environments (at least, best for the administrator), the admin is accorded considerable scope and flexibility concerning how to support the others in the workgroup, and the person to whom the admin reports (thus, the de facto "boss") will assist him or her in getting involved with projects. The admin may perform research, for example, to assist with a particular project. Here is the story of how a young woman made her admin job into something quite different.

Mandy started as an admin with a software products group. Almost immediately, her boss took on a new responsibility, which opened the door for Mandy to do the same.

> I took the admin job working for a director. Shortly thereafter, he became the director of marketing for the networking group, at which time I started to do various projects outside of the admin role because I wanted more work and to grow my job.

Her manager welcomed her offer to expand her role, and she took on additional tasks.

> One of them was working on a solutions guide. What I did is start to create the solutions guide of all the third-party products that were created for our product area. This involved working for all the ISVs [independent software vendors] to get their product listings, information on their company, and to create an actual guide containing all of this information. This job required a lot of coordination—working with a design firm, creating the document itself, and then working with a contractor to actually print the document.

Another woman (and yes, there are male administrators, too!) had been an admin for only three months when she made these observations about the position:

> I think what I like the best about being an admin is that you get to know a lot of people in a lot of different positions and they get to know you. You can also take the position and grow it if you show that you're competent.

Her manager wanted someone who, above all else, had the right attitude. He struck a deal with her when she was hired: If she performed her regular job well, he'd help her develop the skills and background to move into a new field after a couple of years.

> His biggest requirement was to have someone with the right attitude, because you can't change that in a person. I really liked how he had the idea that an admin can go somewhere. As an admin, at least I'm in a place where I can see different types of positions and what they do. You build relationships with people who are willing to talk to you. It's kind of like informational interviewing while you're on the job.

The major dividing line between the admin and the other members of the work group is that the former is an hourly, nonexempt employee—not exempt, that is, from the laws governing work hours, overtime, etc. Another difference is that admins are located in work spaces that look different from the others'; it is rare for an admin to have a closed office, for example. The position has served some people very well—admin jobs can pay up to $55,000 or more a year, in some cases—and some admins are very pleased with what they are doing. But some are not, and when an entry-level job opens up within a professional category, several admins usually try for the position.

The issue of status in the admin job, like that of the secretary in other industries, has provoked some heated controversy. Some admins feel that they are overqualified—for example, by virtue of education—for the job and that they deserve to be treated with more "respect" and be helped to move out of the job into an entry-level salaried job as soon as possible. But good admins are hard to find, and a person who may feel overqualified may get locked into the position if he or she does a good job. I interviewed some people for this book who started as admins and who became professionals very quickly; I've also known some admins who quit in frustration. The solution is to have an understanding with your manager when you start out as an admin: You'll do a great job for a couple of years, and your manager will help you learn and grow, and will sponsor your move to a different position.

Business Development (Market Development)

In some high-tech companies business development means looking out for acquisitions—of technology or, perhaps, of an entire company—but in the present case it's much like a sales job, except the focus is not on quarterly results but on longer-term prospects. The business development manager is expected to become an authority on a particular segment of the market, identifying likely targets for future sales. For that reason the title is sometimes given as "market development." The experience and skills required for getting into a business development position are, like those for sales, extremely varied. At some point the prospective business deal is turned over to a salesperson, and the business development person goes on to look for and develop other opportunities.

The business or market development function is often organized along industry-specific lines. A person specializing in the financial markets, insurance, banking, and so forth, wouldn't be expected to be able to switch to agriculture or heavy industry. Anyone doing market development within a particular industry rapidly builds up an extensive list of contacts; this can be a lifeline should things get tough in the high-tech company and there's a period of layoffs. Either the company keeps the business development manager precisely because it doesn't wish to lose those contacts, or those contacts become a huge selling point for the manager's next position in another company.

These jobs can be fun and involve travel and dealing with potential customers. There's a little less tension than the regular salespeople have to bear, though when the quarterly numbers are in jeopardy, it is not unknown for a sales VP to enlist the business development types as sales reps.

Channels Management

Traditionally, channels are independent companies that the high-tech manufacturer contracts with to help get its products to the end users. Such channels can be of various types: master distributors, distributors, systems integrators (SIs), telesales, value-added resellers (VARs), and original equipment manufacturers (OEMs). While other industries also use sales channels, high-tech channels are affected by rapid changes in the marketplace and in the technology, just as are the producers of hardware and software themselves. This puts a premium on good management of the channels. In most companies the channels management function comes under sales. Many nontechnical people can be found doing channels-related work in the high-tech industry. The high-tech industry relies on sales channels to get a "multiplier effect," to multiply the number of salespersons representing its products; it also uses them to implement marketing programs and to sell into situations where customers want a mix of products and services. Where the channel has some special expertise—say, in the oil industry—the high-tech manufacturer can use the channel to achieve credibility in that industry. Channels are usually required to carry inventory, relieving the manufacturer of this burden. Services, too, such as repairs, spare parts, first-line service, and customer training, are often provided through channels. The creation and maintenance of an effective channels system is increasingly a major part of the activity of most large high-tech organizations. Asking someone inside the high-tech company about its channels organization may yield the information that the organization is scheduled to expand.

Many discrete activities come under the general rubric of "channels management." These include channels communication, recruitment, development, sales support, sales training, and technical training. We'll look at some of these functions in depth in a bit.

What is the importance of channels in the high-tech industry? Some high-tech companies sell directly to customers because their products are very expensive and low volume, so it's worth having an in-house sales force. The same may be true when products are relatively inexpensive, and a potential customer may be considering buying thousands of them.

Having products sold by an in-house group has some advantages. For one thing, the company has more control over the whole complex sales process. Selling high-tech products isn't just a matter of providing some brochures, putting on a presentation, quoting a price, and asking for the order. It's a process that can last for months, and one that may involve dealing with many different levels and functions within the customer's organization MIS (management information systems) directors, financial officers, purchasing agents, technical staff. The competitors aren't sitting back, either. Each of the companies is having a go at the sales prospect, boosting its products to the customer and spreading FUD (fear, uncertainty, and doubt) about the others' products. So for big deals most high-tech companies would prefer to exercise as much control over the selling process as possible.

But in the real world control over all aspects of sales often isn't possible. Customers are many and spread all over the country—indeed, all over the globe. Direct sales are very expensive. Many high-tech companies just can't afford to have the coverage, in terms of "feet on the street," that is necessary to be successful. The independent companies that act as sales channels offer the high-tech company the opportunity to vastly increase the number of sales personnel selling the company's products because such channels have their own sales forces. Furthermore, even if a high-tech company has large numbers of really good salespeople on board, all of whom are well versed in the products, it is simply impossible for the sales staff to know all they would have to know about the special needs of all their potential customers. Customers for high-tech products are in areas as complex and diverse as railroad operating, forestry management, tax administration, academic administration, and telephone network management. Such customers expect potential suppliers to have more than a layman's understanding of their particular industry. They will only do business with salespeople who demonstrate a real knowledge of the problems they're trying to solve. Some channels do have special understanding of certain market segments, and this is an efficient way for companies to sell into these segments without having to become very knowledgeable about their special needs.

This channels system has to be managed by people who have good business sense, who can communicate effectively by telephone, fax, e-mail, and in person. Channels have to be well managed because they are now representing the company, and to the degree they are used, the company's fate is in their hands. A technical degree is not necessary to work in channels management, but you must know the company's product line, plans for new products, and have excellent business sense.

Here are the major types of channels used by the high-tech industry:

Master Distributors. One of the worst things a high-tech manufacturer can have is an inventory of unsold product on hand. It costs money, requires storage and insurance, and is demoralizing to think about. Having a distribution channel in place solves this and many other problems (although, like anything in this life, there are tradeoffs). The master distributor purchases in large quantities, improving manufacturing efficiency, and, in turn, sells your product within a defined geographic area to resellers of various kinds, including VARs. The master distributor is usually required to provide the high-tech manufacturer with a "rolling forecast" of sales over the next six months, so the company can plan its requirements for parts and schedule its factories' operations efficiently.

To give you a picture of some of the problems that nontechnical people get involved with in managing distributors, consider first the ideal situation, which has never existed and will never exist, but is still the dream of every channels manager. Ideally the distributor's credit is superb, so there is no question that the company will get paid; the distributor has a staff of well-trained, highly motivated salespeople; and because of the perfect mutual understanding that underlies the relationship between the high-tech manufacturer and distributor, there is no danger that the manufacturer will find itself competing with its own distributor for a large chunk of business.

Distributors are limited by contract to certain geographical areas—"geos," in the high-tech lingo. In the ideal situation, the distributor would rather deliver his or her first-born son personally to Pharaoh than ship even a cable outside his or her agreed territory of the western United States to, say, Tallahassee, or maybe Taipei.

Ideally the master distributor is so happy with the deal struck with the high-tech company that taking on a competitive line of equipment is out of the question. Or if forced to do so by market circumstances, at least the distributor would set up a distinctive company to handle the competition's useless stuff. Finally, the distributor would be well-funded and would insist on bearing a fair share of the costs of trade shows, advertising, and the like.

Unfortunately, this ideal situation never really exists. In real life the ideal is what you work toward but never quite achieve. Managing distributors, as with many other areas of business, calls for a delicate balance of toughness and diplomacy. The distributor is in a kind of partnership with the high-tech company, but the distributor may (and almost always does) have many lines of products it is carrying, some of which may be directly competitive with the high-tech company. The high-tech company has placed all, or many, of its eggs in the distributor's basket. If the relationship breaks down, it can be very bad for the manufacturer, who may see its sales slump or even cease for a period of time in a given geo, until a new distribution channel can be set up.

OEMs. OEM stands for "original equipment manufacturer," a company that buys a high-tech company's product and incorporates it into something that they make. The OEM then sells that product to an end user. An example is a flight simulator, which is how airlines train pilots without risking their airplanes at the same time. Flight simulators have computers in them, but no one sees the computers except the service personnel, who may occasionally have to take a look and give them a squirt of oil. (Just kidding. You don't really oil them.) Another example is your telephone company, which may at this moment be bidding to upgrade the phone system of some developing country. Your phone company may be bidding a switching system that contains computers, but no one knows or cares about the computers as long as the calls get through. Some of these OEMs may be giants such as AT&T; others may be small start-ups with a great new technology they are just starting to bring to market. Sometimes the company selling to the OEM may prefer to have all contacts be between the OEM's regular sales force and the customer, or even between a dedicated salesperson and the customer ("dedicated" because the customer needs someone who can understand its own complex product, and this requires a specialized salesperson who attends only to this particular customer).

Systems Integrators. In order to understand what a systems integrator does, let's consider an international business situation based on an actual case. A high-tech company gets wind of what looks like a good sales opportunity. The potential customer is the Ministry of Finance of Transbaltania. Transbaltanians do a lot of importing, and the citizens are tired of seeing underpaid but heavily bribed customs agents driving luxury cars; the importers are tired of paying bribes and having to get eighteen signatures on a document to release a shipment of canned tomatoes; the docks are crammed with rotting cases whose contents are a mystery. The Ministry of Finance, which has responsibility for customs administration, has decided that one way to get things under control is to computerize customs operations. The estimated size of the hardware part of the deal (and that's what the high-tech company is interested in) is over $3 million. Best of all, though Transbaltania is fourth from the bottom on the list of the world's poorest countries, if the company wins this deal, it is certain to be paid, and not in kropecks, the local currency, but in dollars, because the entire deal is financed by the World Bank.

The company probably doesn't have any representation in Transbaltania, however, or indeed anywhere within five thousand miles of there. Also, such bids usually require other specialized equipment and software that this particular high-tech vendor does not produce and is not in a position to acquire.

In large deals like these (and not just internationally, of course), the systems integrators come in. These companies are sales channels that bid on the overall project, working with a number of manufacturers, each of whom is concerned only with supplying a piece of the total package. Because it's their business to know these things, systems integrators may have been acquiring knowledge about the project for months or even years before the bid documents hit the street, so they're better prepared to win the project than any of the individual companies whose products they will integrate into the overall solution for the customer. Normally, dealing effectively with a systems integrator to win a large contract requires that the high-tech company set up an ad hoc working team; this team may have engineers, and sales and financial people on it. Such teams can be effectively organized and managed by nontechnical people. Because there is always a time constraint to putting the deal together, a premium attaches to effectively "driving the process." Participation in such a team is a good way for the nontechnical person in channels management to acquire experience in winning large chunks of business, and this always gets the favorable attention of senior management.

VARs. The term *VAR* stands for "value added reseller." These are companies who buy a product, say, a computer, add something to it, and resell the package to the end user. The difference between an OEM and a VAR is that the VAR sells the computer with the manufacturer's name still on it, and everyone can see that it's a computer (or Web phone or set-top box or whatever).

What is it that the VAR adds? It can be a piece of specialized software that enables a specialized need to be met. Let's say that the VAR has developed some software that enables the chocolate manufacturing process to become more efficient by a factor of 10 percent, by more accurately controlling the quantity of chocolate liquor squeezed from cacao nibs. Chocolate makers from Brussels, Belgium, to Hershey, Pennsylvania are beating his door down. He decides to sell a company's computers preloaded with his software. Yes, he's like an independent software vendor (*ISV*), but he has the inside contacts with a very specific segment of industry and decides that he can sell a complete solution. This is a neat way for the computer manufacturer to sell into the chocolate industry.

VARs need a lot of support, as well as a system of tracking, because they are usually rather distant from the primary manufacturer—they buy from a distributor who in turn buys from the company. The company must set up a system to monitor their performance and keep track of how much business they are doing; channels management folks periodically carry out business reviews for this purpose. But it's just as important for the VARs to catch some of the spirit of the high-tech company they ultimately represent. Conferences, "kickoff meetings," technical and sales training sessions, and "product launches" are all means of helping to build a feeling of partnership. These sorts of events are partly business and partly like a reunion of an extended family, with considerable attention given to creating a personal team atmosphere. In this way channels management organizations try to bring about more contact between the manufacturer and the VAR, without upsetting the primary VAR-distributor relationship. The requirements for this kind of work are an interest in business and an interest in people.

Let's take a closer look at some of the channels-related positions held by low-tech people:

Channels Development—Here's a description of this position, in the words of the low-tech young man who held it in a large, diversified hardware company. When this interview took place, he had been in a new position for a few months.

> I actually held three jobs, all within our channels marketing group. First I was a new channels guy, looking into all the new channels. Then I had to develop an e-commerce strategy—doing business on the Internet. And then I had to create a general Internet strategy. I guess if you roll it together, the type of work I was doing was channels strategy. I was specifically in charge of North American channels. My job basically entailed looking at new channels of distribution for us and determining if and how we should enter these channels, and specifically helping product lines accomplish their own distribution strategy to enter those new channels.
> You need to be a good salesman. I found that I had to sell my ideas and plans to people within my own company. I had to sell to the sales force, to every marketing group, to the whole division. And of course, I also had to sell to the executives.
> I had to be pretty quick on my feet, and be able to analyze a problem very quickly, determine who my contacts had to be and exactly how I was going to approach them. Other skills are also important. I had to look at what our competitors were doing. The most important thing is to have sharp, critical thinking skills. I think anyone with such skills can pick up the tools, like statistics, that you might use.

Channels Training—Sales channels—OEMs, VARs, systems integrators, and so forth—don't know your company's products as well as your company's salespeople and engineers do. Sales channels need to be trained on the products, and also on all the sales messages—the "value propositions"—inherent in the products. Larger companies have individuals or units whose task it is to keep the channels up to speed on all this. Here's how one man described his channels training responsibilities:

My responsibility is to develop all of the courseware to train our channel partners, and that development includes everything from deciding what classes need to be developed up through developing the course materials. I have responsibility for making sure the classes are deliverable to a worldwide audience. I don't actually teach the courses, but make sure they are deliverable.

I'd say one of the most important skills for my job is . . . the ability to listen and to negotiate, and to have good interpersonal relationships, where you develop a level of trust and where people understand where you are coming from and that you're going to do what you say you're going to do. So, it's what I would call basic business skills that would be applicable to any area of business from a hot dog stand on up.

Contract Administration

This job sounds boring, doesn't it? Paper pushing, filing, yawning from time to time. Yet the terms governing sales, service, and financial relationships between high-tech companies and their most important customers are not left to the whims of a sales rep. They are all carefully defined by contract. At least, an attempt is made to define them carefully. Lawyers work hard to get the perfect contract, which means, of course, perfect from the point of view of the seller, the high-tech enterprise. With luck, the customer will sign the contract as is, but that kind of luck is usually reserved for people who win the lottery. More often, the customer has its own ideas about what will make the relationship fly. Companies such as Boeing, Exxon, PepsiCo, BellSouth, and American Express have lawyers of their own, after all. By the time these folks get through with the basic agreement, it may have all kinds of special terms written into it.

The contract administrator's job is to know the terms of the contract cold and to manage the execution of the contract. The relations established between the parties by a contract aren't static; a contract usually calls for each side to take various actions over a period of time and provides for remedies should things start to go awry. An example is the area of product warranties. Customers want the products they buy to be trouble free, and if there is trouble, they want to know that the manufacturer or vendor or both will stand behind the product. In the high-tech world, new products come out every year and a half or sooner, and they are immensely complex.

Things do tend to go wrong, and sometimes there is an issue of whether a particular case is covered, and what the contractual remedy is. The contract administrator can advise the sales and service people what the contract has provided, sometimes playing a direct role in negotiating disagreements. To do this he or she may have to visit the customer for discussions; set up meetings involving technical, legal, or business personnel; and sometimes recommend that the contract be rewritten to take account of new business realities.

Contracts usually have an expiration date, and the administrator also keeps track of this date and sends out the proper notices for renewal, where that is desired. Other people in the company have to be informed of the termination of a contract; it is awkward for the contract to have expired (or worse, been intentionally terminated by the parties) and to have business continue as before, carried on by unwitting salespeople or other personnel.

Contract administrators often work closely with lawyers, and since lawyers, like everyone else in high tech, have too much to do, they may encourage the contract administrator to do a considerable amount of the work involved in preparing the contract itself—drafting the sections on pricing, product description, and other sections. (This drifts close to, and sometimes over into, the realm of the contract negotiator.) It's a very good thing to win this kind of trust from a high-tech lawyer, as a warm recommendation from the lawyer can be very helpful in the move to the next position.

Contract Negotiation

This position involves negotiating contracts face to face with the customer. You might wonder why this isn't done by a lawyer; after all, a contract is a legal document. In some companies lawyers do, indeed, negotiate the agreements; in others, the contract negotiator, who is thoroughly familiar with the terms of the contract and how much flexibility exists with respect to easing up on some of the contractual requirements, will do much of the front-end work. The legal staff serves to back up the contract negotiator and may be called in only when almost everything has been agreed to between the parties except for two or three crucial points.

In a sense, then, a contract negotiator is somewhat like a paralegal in a lawyer's office. The difference is that the negotiator is dealing constantly with the same form of contract (or contracts) and gets to know their provisions extremely well, including the business and legal rationale behind every phrase. The negotiator would not, however, draft a new form of contract. This would be done by the lawyer.

The principal requirements for moving into such a position are an excellent command of English, the ability and desire to read and understand complex documents, good interpersonal and negotiating skills, and a high-level knowledge of the company's products and business model. It's possible to move from a clerical position into a position as a contract negotiator, if you've won the confidence of the legal group. It helps to have made friends with a member of that group—in other words, to have a sponsor. You may start off working with rather simple agreements—for example, confidential disclosure agreements, which must be signed by anyone to whom your company wants to reveal confidential information. Then you move on from there and, eventually, end up working with a sales rep when a deal is about to close. Travel is involved, so the job can be rewarding in that sense. Sometimes contract negotiators develop a desire to get a law degree and "move up the ladder." Others develop a taste for sales and go in that direction. It's an interesting job, and perfectly respectable.

Corporate Communications Manager

This position combines elements of public relations and media relations activities that are covered later in this chapter. Here's a description in the words of a young woman who works for a start-up:

> I double-majored in English and communication. Now I work for an Internet company as corporate communication manager. So I'm in charge of all the press releases; I'm also in charge of writing out technical manuals, which involves talking to the developers and engineers and taking what they're developing and translating it into language that laymen and decision makers in finance departments and places like that can understand. I'm kind of a translator from the technical world to the rest of the world.

Other things this woman does include writing the company newsletter. What's interesting is that so many different activities are combined in one position. Of course, that's because the company is a start-up, with about forty employees. As you'll find out in chapter 8, having to be a Jack or Jill of all trades is not unusual in the world of start-ups, and many people find this to be one of their principal attractions.

Just before this manuscript was finished, I heard that this particular start-up was running out of gas, as so many do. The young woman, however, had built up a strong resume, and soon had offers from three other start-ups.

Customer Response Center

As with some of the other service functions, working in the response center involves dealing with people over the telephone who are calling you with problems or questions about your company's products. The response center, in this case, could be located in CS, but could also be in the marketing or sales departments. These calls usually don't involve complaints. Instead these are people seeking information. Sometimes the questions get into quite technical areas, but the joy of this job is that the caller almost never expects a quick answer. You can concentrate on getting the question down correctly, getting the name of the caller and the phone number, and committing to call back within an agreed time with at least an interim response. Then you get the answer from the right person.

As in telesales and in the customer service function, you deal with customers or partners outside the company. It is very helpful to have experience involving customer contact on your resume. Customers really are the most important people in the world when you're in business, and business people, including hiring managers, value people who have dealt with customers successfully. In a customer response center position you learn who and where the internal technical resources are in the company, and you build relationships with these people. You also build your technical understanding day by day, though this may be imperceptible at the start. You may think you are just relaying information blindly from one person to another, but the day will come when you will suddenly realize that you know the answer to a caller's question. You won't have to go bother your friend in product engineering, as nice as he or she has been about it up to now. In a few months you will know a lot, and you'll know that you know it. Time to update the resume again by including the technical knowledge and people skills you've been building up.

Sometimes people outside the high-tech industry hesitate to get involved with any job involving the telephone. It seems to be too low in status. Well, listen to the words of a woman who's been doing this kind of work for a few months at a software company, who got her position through an agency:

> Actually I wasn't interested in the position, but they encouraged me to come here and interview for the position because I might change my mind. And they turned out to be right. Once I found out more about what I'd be doing and I met the people, I thought maybe it would work out for me in the long run. At first I thought I was going to be answering questions eight hours a day. It turns out that that's about 25 percent of the time. A lot of the callers ask me how something works. I don't always know, but I'm reading all the spec sheets and trying to understand as much as possible. And because I'm hearing it so much, I'm learning it. And if no one in the group knows the answer, I have to go to an engineer. It'll take a while because I haven't had any formal training.
>
> I've had a chance to learn about the workstation and the applications, and that's very useful anywhere. I can allocate an hour or two a day to myself for learning, and I'm doing some other projects as well.

Here's another person talking about her job managing a small customer response center (here referred to as a "call center"):

> My current job is managing a small call center. . . . We receive incoming calls from customers who need assistance in getting directed through the company, if you will. So we're basically trying to . . . provide the customer with the right resource to get their final answer. This is not really a technical position—it's more general—but we are certainly focused on a lot of applications in order to do our job. So you still have to have an understanding of the technology, what's out there, what's happening with the Web, with software applications, and so forth.
>
> Most of the people coming into this organization are

first-time workers [in high tech]. They will usually not have any background in high tech or our product line. They normally come straight out of school, or sometimes they're still in school. We're open from six in the morning until five at night. It makes a good place for employees to be while they are attending school because they do have some flexibility with hours.

An important skill is time management . . . but probably even more important are the customer-interface skills. I tend to get the customers who are difficult. And it's very important that we defuse those situations and get the customer a resolution.

For potential hires I try to find people who'll be comfortable online, taking multiple types of calls, dealing with ambiguity—because you never know what the next call is going to present. . . . Multitasking is important because we're requiring people to not only talk to the customer but also to research information online while they have the customer on the phone.

We give everyone a one-week training period, starting with a history of the company, so they have some background. We teach phone etiquette, how to deal with a demanding customer, and we actually do practice calls from outside to let them have the opportunity to make their mistakes. And we don't let people fly solo the first day out of training; the first day on the floor they're with a mentor, who's listening in with them and helping them, so there aren't any problems they can't solve then and there.

Telephone jobs are good springboards to other positions in the high-tech company and are natural places to start growing your job using the techniques described in chapter 7.

Customer Service Business Manager

This person runs the profit-oriented customer service function as an independent business within the high-tech company. The position is usually very well paid, and may be at the director level. You must have several years of experience inside the CS group to move into this position. Ideally the CS business manager would also have a business degree (which one can often acquire at company expense), but it is not necessary to have an education in technology; you can acquire the requisite knowledge along the way.

The CS business manager usually has a number of technical people reporting to him or her. Because of a lack of adequate staffing, customer service operations are often highly tiered, to provide several layers to which problems can be "escalated." The business manager is constantly preoccupied with how to handle the workload, and if you, as an admin, marketing person, or temp worker, can relieve any of the load, you'll be assured of your manager's gratitude and support.

Whether rightly or wrongly, in many companies technical service people have the reputation of being cantankerous and hard to manage, and the CS department often deals with customers who are unhappy because of system, software, or network problems. The best CS managers, therefore, have excellent people skills, as well as an in-depth knowledge of a very specialized business.

Customer Service Representative

The customer service (CS) function has traditionally been one of the main channels by which nontechnical people have moved into and up in the services that the high-tech industry. Customer service includes all the services that the high-tech company provides, or is prepared to provide, after the sale of its products. In some companies the list of services for hardware and software is long, including warranty service, sale of upgrades and spare parts, service contracts, training, consulting, hot lines, and, when the company decides to contract some of these functions to an outside contractor, the management of third parties.

Customer service is a function that often has to fight hard for headcount, marketing funds, and other support. That's because it isn't seen by many decision makers as an important source of revenue. But good customer service is a very important part of sales. Excellent service opens the door for repeat business—the best kind—so the quality of people in the CS organization is crucial. As one manager told me:

> If you look at a composite set of skills, people doing customer support in a technical environment must have a much broader skill set than someone sitting in a development role developing the core product. Because not only do they need to understand the technical aspects, and how that translates into actual usage—which many development engineers couldn't do if their

life depended on it—but they also have to be able to communicate the findings and information to a customer.

So they have to have, in some respects, the communication skills of a salesman, the technical skills of an engineer, and the ability to deal with the customer like a psychologist. Finding people who have those skills in a more-or-less equal mix is an extremely difficult task. That's why you see so many tiered and layered support organizations—because you can't find enough people with those skills.

It's been said that the sales department lives with the customer only up to the moment of the sale; the customer service department lives with the customer forever. Certainly a large part of a high-tech company's reputation in the market depends on the market's perception of the range and quality of its service. In fact, customer service is rapidly becoming an important market differentiator between companies; because of this, high-tech companies are investing more in this area, which means there are better chances of job openings.

A senior customer service manager was asked about job opportunities for people with no technical degrees in the customer service field. He said:

> You can certainly find nontechnical people in customer service. I'll give you a perfect example. Some years ago I was running a training center, and the big problem we were having was that we couldn't find enough technically trained people, with college degrees or whatever, to fill the rapid growth that was needed in the support organization. Now, these are technical support functions. So what the corporation wound up doing was to build their own technical people. We hired people across the country with liberal arts degrees, and we put them through a six-month training program on the particular technology and the customer environments where they were going to be providing support.
>
> The amazing thing about this strategy was that, in evaluating the customer service department five years later, the most effective support people proved to be the ones who came with a liberal arts background. The reason for that was communications skills. People with a purely technical education many times miss what is a key component in customer support, and that's the ability to communicate with the customer.

If all these comments from the experts are true, how does a nontechnical person get started in what is, after all, a "technical support" function? There are a few routes. One is to start by working on a hot line, where you field questions from the customers and try to find out the answers. As I've stated elsewhere in this book, many job seekers tend to shun such jobs. That's too bad, because *any* job is the right first job in the high-tech industry. After you get in, you can make your moves, and in the case of customer service, the right moves might very well be getting the company to train you, at its expense, in technology. Other routes into customer service could be via a marketing or operations function. Support organizations in larger companies have people providing marketing and other services, since these organizations are generally run as profit centers. Finally, a possible route is by starting as an admin and then sliding over to another CS function in a year or so.

Though larger companies appear to have resolved the question for themselves, there is still a debate going on in the industry as a whole to whether service should be a cost center or a profit center. Being a cost center means that everything (or almost everything) that the CS department does is subsidized by the sales function. For example, the sales department might be required to kick in some percentage of its revenues to fund after-sales service. Being a profit center means that the service department is run like an independent business. The company has invested in it and wants a good return on its investment—as good a return as from any other part of the company.

It's always better to be part of a CS group that is run like an independent business because such a group charges directly for its services and tends to be sharper and more responsive; also you are very close to the customer. (This positioning is excellent for learning and for survival when times get tough and "excess" employees are at risk.) CS groups that are well-managed tend to have a higher level of esprit de corps, perhaps because in many companies, as you may have deduced from the comments of the managers quoted above, they are understaffed and overworked, and depend on their morale and energy to win the day. CS is a sort of U.S. Marine Corps of high tech; everyone depends on them and at the same time beats up on them. As in the USMC, the level of professionalism in CS tends to be very high. A customer support rep said:

> One of the payoffs for career support people is that they resolve customer problems. You're faced with a situation where the customer's unhappy and the product's broken. You take all those factors and turn them around into a plus.

When looking into a customer service job, you should inquire of the hiring manager whether the CS operation is run as a profit or cost center. Just your asking the question will impress him.

Customer Training

This function is usually to be found in the division of the high-tech company that deals with services of all kinds: consulting, repair ("bug fixes," in the software world), and education. It involves training customers who buy a company's technologies, or products, in how to use or maintain them. Some large companies who have in-house training and education departments will offer one set of courses to customers, and another to their own employees. These dual sets of courses may cover very similar topics, but because of a desire to keep some aspects of a product or process confidential or for other reasons, there will be some differences between them. "User" courses are usually nontechnical, and training departments don't require a technical background of the people presenting them. Sometimes customer courses can be very technical indeed. Quite often these may be offered at the customer's site.

What is important is having "platform" (teaching) skills, being articulate, and being able to present the material in an interesting way. Almost anyone with these skills, who has also acquired a good command of the company's products or technologies, can be taken seriously as a potential trainer.

An allied activity is that of course developer; degree and certificate programs are available in course development, and nontechnical people who meet the basic qualifications for being a trainer can move to the developer position rather easily by taking an after-hours program.

High-tech companies usually try to present such courses in their own facilities, simply because that is the easiest way of satisfying the equipment requirements; user courses require one computer for each student and sometimes a local area network, servers, overhead projectors, audiovisual equipment, and so forth. The companies may also rely to a degree on outside contractors (individuals) for presenting courses.

Forecasting (Orders and Shipments)

Forecasting is the means by which companies seek to synchronize the supply of goods with the market demand for those goods. Usually, a small group of persons is charged with this responsibility.

If one is selling toothpaste, synchronizing production with demand isn't much of a problem, since both the characteristics of the product and the demand for it remain fairly stable over time; you can look at historical sales numbers and pretty well predict the future. If one is selling a high-tech product—a workstation, for example—the product may have a life cycle of only a few months before a newer product moves into the same market space. This complicates the forecasting task because historical sales numbers may be no guide at all to what is going to happen in the future. High-tech companies rely to some degree on the past, but they rely even more on what their customers say they want and plan to buy. Customers may inform the company that they need a workstation that sells for a certain price, and if they can buy them at that price, they will buy hundreds or thousands. If they can't, then they might decide to buy PCs at a much cheaper price.

If this information is obtained from a large number of reliable customers and the manufacturer is planning to come out with a workstation at the right price, estimated sales volumes can be deduced. A member of a forecasting group describes how this process works in practice:

> We would look at a few things. One was historically what we'd been selling at various price points. We divide our product line into four different price points for simplification. We look at requirements as reported by our sales force. And then we look at our own product rollout plan, talking to engineering and operations people. It's very much an art because of the kind of business we're in. It requires a quite extensive understanding of the product. There's a lot of uncertainty in it.

Forecasting can be critical to the success of the company when there are shortages in the components used to make the product, causing long lead times for ordering the product. A few years ago a major computer company underestimated demand for a new machine and, as a result of a shortage of monitors from a supplier, found itself with a huge backlog of orders. Luckily for the company most of its customers were willing to wait for some months before getting the systems. The risk is that customers will become angry at the delay and buy someone else's systems.

You don't need a technical degree to be a good forecaster, but you do need to know your company's product line and understand the dynamics of the market. Companies tend to favor people with business degrees for positions such as forecasting. People in forecasting get good exposure to the engineering, sales, and marketing departments, and to top management. If forecasts are significantly off, factories will be making the wrong product mix, or too few or too many of a given product; the purchasing department may likewise under- or overestimate the company's requirements. So forecasting is an important assignment.

Human Resources

Probably no professional area has been subject to so much criticism and experienced so much turmoil in recent years as has the human resources (HR) function. The reasons for this are partly due to the way this function has usually been staffed, particularly at the lower levels, and partly due to the economic squeeze that has been placed on many large high-tech companies by declining profit margins in the hardware business. Yet HR is a critical function within a company, since employees' salaries and benefits usually account for the biggest single cost in the high-tech industry. Job turnover and training of new hires add significantly to these costs. HR can offer a real career challenge to articulate nontechnical people with training and experience in effective communication, psychology, educational management, and other fields. But before considering a career in high-tech HR, you should be aware of some recent history.

The human resources area in the high-tech industry was traditionally an area in which admins, secretaries, and clerical staff had a chance to move onto the first step of the salaried employee ladder. "HR generalist" was the generic title for such entry-level positions, which dealt with the administration of benefits and compensation plans, tracking of employee records, and so forth. These sorts of activities deal with the "technical" part of the HR function. They don't require a great deal of creativity or higher education, so they became positions highly desired by low-level hourly employees. One of the problems this situation engendered was that, as companies grew rapidly in the 1980s, HR departments also grew, and following the general rule of promoting from within, many HR departments promoted people out of the "technical" side of the function into managerial and professional positions. Some of the people thus promoted had already acquired the requisite training by attending courses and seminars and were able to make the transition successfully. Many, however, simply found themselves out of their element—promoted too high, with decision-making power and with responsibility for managing others but with no training in how to perform these functions.

A senior HR manager told me:

> I told my staff when I arrived here, "Look, HR has been this dumping ground. If you've got an admin who's at the top of the range and you want to give her more money, you tend to put her into HR. Because after all, anybody can do that job; there's no technical knowledge associated with the field as a profession. I think that's wrong. When I go home and I talk to my parents about doctors, it conjures up an image for them of this person with an understanding and knowledge and techniques and capabilities to save lives. When I say the same thing about attorneys or plumbers, those conjure up similar notions. When I ask them about HR, they think there's nothing to HR, that it's just common sense. And they're wrong. There's a whole lot of knowledge and skills that we can bring to the party. So I'm trying to articulate those very, very clearly and say this is what I expect people to have as a journeyman's card before I'm going to allow them to practice this profession within this organization.
>
> Now, I don't have these parameters clearly identified yet. Once I do though, I don't care where those people come from—an admin, a graduate student. [But] they have to demonstrate that they do have those abilities and skills in order to be able to perform at a minimum level of competence. . . . Because the fact is, HR people mess with people's lives. And I think there ought to be a certain level of competence that somebody has to display before I'm willing to let them go and mess with somebody's life.

In an industry in which top managers were prone to make statements such as "our employees are our most important assets" or "the only real asset this company has is its workforce," and in which the HR department was putatively in charge of employee welfare, a gap rapidly grew up between employees' expectations of HR and what the department was able to provide. A typical comment from an employee describes the situation:

> A minor part of their job should be administration, yet that seems to be a major part of their program. They don't have consulting skills. They don't have change-management skills.

A vice president of HR observed:

> I wouldn't say HR has evolved, but it is evolving. If you talk to any HR person around the country, what they'll tell you is that we're trying to get out of the [routine clerical] work. We are trying to get into more strategic issues associated with running companies. We're trying to add more value, and we're trying to be able to document and demonstrate that we can provide a competitive advantage to the company. The question becomes, how do you demonstrate this? If we can't [provide a competitive advantage], then we're not really serving the company well, and we might as well go back to the old process—the papers and all that sort of crap. We all agree that needs to be done, but that's not where our value add is in the organization.

In its new approach HR often finds itself battling line and staff managers who have never been part of an organization that had a good human resources function. So their expectations are low, and therefore it's been relatively easy for HR people to meet those expectations. In its new function, HR is trying to be seen as a coach and counselor—a true support function for the business as a whole.

HR includes a very important area usually called "employee communications." This encompasses everything from seeing that top management decisions are conveyed to the workforce through announcements, newsletters, and the like, to providing a channel for airing of grievances, including possible violations of employees' rights. Another important HR area is employee training and development. Training departments usually come under the HR umbrella (although training is considered separately in this chapter). Administration of benefits and compensation are HR specialties, as well.

These days, the top companies aren't very interested in hiring people into HR unless they've got a master's degree in human resource management or a closely related subject. It is possible, though, to try for an internship in such a company or for a full HR position in a medium- or smaller-size company. Smaller companies will tend to be more open to good candidates who don't have graduate degrees. Perhaps this is open-mindedness, or perhaps it's because such people cost less than a person with a master's or a Ph.D. Whatever the reason, the field is very important, and extremely interesting as a career.

ISV Relations

ISV stands for "independent software vendor." ISVs are important because they create the applications software that businesses actually use to solve problems. To track accounts receivable, inventory, and payroll, for example, large retail chains use software that some ISV developed and that they bought, maybe directly from the developer (the ISV), or maybe through a systems integrator that managed the installation of their total information management system. Or maybe they bought it as a package deal with the hardware. It's a fact of life that a computer without applications that will run on it is just a useless pile of parts. That being the case, it's awfully nice for the retailer to know that the software he bought to run his business is compatible with the hardware and its operating system.

A person working in ISV relations for a company making operating systems (see chapter 5) can either be like an account manager, making sure that the company is responsive to the concerns, issues, and needs of certain very important software developers, or like a program manager, devising and implementing programs to make it easy for all developers to write applications to the particular OS.

There are tens of thousands of developers out there developing applications software, some in small one- or two-person firms and some with hundreds of employees. The computer companies that develop operating systems pay them a lot of attention because they want all that software to run on their computers, or no one will buy their computers. So they try to work closely with the software developers to get them to develop their software on, or "port" it to, the companies' particular machines (usually referred to as "platforms"). Microsoft, a giant in the software industry, owns the ubiquitous Windows 95 and Windows 98 operating systems, as well as the new NT operating system. Microsoft has no trouble attracting most ISVs for the very simple reason that just about every PC in the world runs on Windows, and that's a big software market for the ISVs to shoot at. But Microsoft takes nothing for granted and offers a full array of programs and other assistance to developers. Hewlett-Packard, Apple, IBM, and Sun Microsystems are examples of other companies with excellent ISV support programs.

Unfortunately, the best support programs aren't of much use when the developer community perceives that you're losing market share, since it no longer pays to write software for your platform. As Apple's market share dropped from 11–12 percent in the early 1990s to around 4 percent in 1998, the developers reluctantly deserted the Mac. The company hopes, with its new products, to win back market share; if that happens, the developers will return, and the stores will again burgeon with Mac software.

A technical background is not necessary to work in this field. People who work in ISV relations work closely with the software applications companies, line them up for industry trade shows (to showcase the software capabilities on the machines), get them to participate in important customer visits, and are the contacts for all sorts of business-related issues. They play key roles in coordinating the efforts to get the ISV to develop or port to the company's operating system, a consummation so devoutly to be wished that, to help out with the port, companies will lend engineers to ISVs (who may have a fabulous piece of software, but be only a tiny outfit of a few people). A port may require manpower and money as well, and for applications deemed critical, a large company may fork over a lot of money to "buy the port."

Marketing

As you can see from the list of activities under the marketing heading, few people seem to be able to define exactly what *marketing* means, yet every nontechnical person—and many technical people—in the high-tech industry seem to want a job doing it. What exactly is marketing in the high-tech context? One person described it thus:

> Marketing is doing everything you have to do to sell the product except actually getting out there and selling it. What that means is knowing who needs the product before it is developed; what the competition has on the market, or is likely to have by the time your company's product appears; the price at which your product could be sold; the effect this product's appearance may have on other products your company makes; how the product should be sold (direct sales force, through channels, over the Internet, etc.); the resources needed to build the product (and whether these can be borne by the price); and so on.

Some high-tech companies are run by engineers who give scant attention to marketing. Here's a story of a whizbang product developed by a team of top engineers at a start-up company, but without the benefit of any input from marketing, that demonstrates the importance of marketing activities before a product is even developed:

> Well, this new product line was considered the hardware designers' dream product. It never got going. It was a classic failure. It derailed the company, big time. You've got to look at the market requirements; you create a specification and you try to build something the market wants. If you leave it to the engineers, they go, "Wow! If I can put this and this together, I can make this, and then everyone will come running to my door!" The better mousetrap thing. No marketing input, no research, just engineers' dreams. It ate up all our venture funding and all the profits.

This little story is not uncommon, unfortunately, and many companies waste money developing great solutions for which there is no problem, or which demand a price no one is willing to pay. One of the important tasks of the marketing organization is to assure that the company is aware of and responds to the needs of the marketplace, not of its technical staff.

Incidentally, the person who recounted the above tale of woe is now the marketing manager of an instrument company, which though small, is now highly profitable. This person's preparation for the world of high tech consisted of an undergraduate degree in psychology, with a minor in philosophy, a love of writing, and an ability to type really fast, "from having to do all those papers."

The traditional business school jargon describes marketing as consisting of "the four Ps" (product, pricing, promotion, and placement). Sometimes a fifth "P" is added—for positioning. In planning new products it's not enough to know what the market wants; as stated above, one must keep an eye on what the competition is doing. Company A will try to "position" its new offerings to edge out the competition in terms of "price-performance," or some other criterion. If the market wants a laptop computer with certain technical characteristics, and company A's competitors have already introduced some in a certain price range, it won't do company A much good to introduce its own laptop at a slightly higher price. Being in marketing can be like playing a game, in which you try to outguess and outplay your competition.

In addition to these activities, marketing people can and should be members of various committees dealing with business policy and tactical issues. An example is the committee or group that reviews and approves pricing proposals for new products. Some high-tech marketing activities, such as advertising, product packaging, and preparation of catalogs, are almost identical to those activities in other types of industry. If there's any difference, it's that it is important for the high-tech marketing people to have more product knowledge than their counterparts in other industries because the features and benefits of high-tech products are more complex than, for example, those of a new soap powder or cereal. So acquiring some real product knowledge is important. But what is most important in marketing is communication. A woman who is a marketing manager in a software division of a large company observed during an interview:

> What's happening in Silicon Valley is that these companies have grown very rapidly and they tend to draw on their engineering staff and put them into marketing or managing situations. And it has been a disaster in many cases. . . . The engineers don't know how to communicate with their public, and they really don't want to deal with people. Your ability to communicate . . . is probably the most important part about marketing, and you're doomed in the role without it.

Brand Management—Brand management is an element of the marketing function that is relatively new to high tech. I asked an executive at a large software company to talk about why his company was making such strenuous efforts to establish their brand name:

> The task is really to create awareness of the brand and associate it with the company. A major part of this is to communicate the value of the brand and to build programs to convey the real content of the brand. So branding is kind of like establishing a promise.

Branding is "establishing a promise." That's the shortest and best description I've ever heard of this function. The brand manager works across the company and with outside ad agencies to convey this promise to end users, distributors, OEMs—the world. It's very helpful if the candidate for a brand manager position has done a stint in an advertising agency, but it's not necessary. In one software company, a woman learned that the brand manager was moving to another position. She started attending meetings with him before he left, and when the day came, she told her manager she wanted to take on the brand management responsibility along with her existing job in marketing communications. She said she was certain she could do the job if she could have an entry-level assistant working for her. Her manager agreed. That's growing your job!

Events Management—This function is usually included in the marketing area. It has to do with all the exhibitions, trade shows, promotional days, and the like that the high-tech company uses to showcase its products. Here's what one events manager had to say about what she did:

> I set up all the logistical arrangements for events . . . finding the venue, working with contractors to arrange electrical signage, staffing the booths, and promoting the event. Then I have to be on site making sure that the public is getting what they're paying money for in the way of contact with the company. So it's a pretty comprehensive role.
>
> To a high-tech company, it is very important because there is so much competition. And it is important for them to have an area where they can arrange for demonstrations of successful applications that the public can see. Because sometimes it's hard for the public to get their brain around some of

this technology and what it does, what problems it's solving, and so on. That's what translates into future sales.

Field Marketing—Field marketing is a part of the overall marketing infrastructure. An international field marketing manager described the position in these words:

> It's really designed to empower the sales force, to provide the tools and the vehicles of communication for them to understand what market opportunities there are, what the strategies are to get there, and what solutions the company is providing to lead to its success. So there's a lot of creation of tools, such as sales binders, product binders, and seminars. We work generally to promote the products in the field.
>
> There are regional field marketing people, who cover different areas of the world. That's necessary, given the fact of different cultures, different stages of technology, and so forth. There's a second group that does much of the actual production work—binders, design work, purchasing of direct mail services. Last, there's a communications group that does a lot of the Web sites, newsletters—people who are very strong in the ability to communicate different ideas and concepts. What field marketing focuses on is the more tactical execution.
>
> To me the best thing about my work is that there's just a fabulous mosaic of peoples and cultures out there. The people are the most interesting part to me. And it's interesting to see how they adopt technology, and how it impacts people's lives.

Field marketing jobs involve travel—whether international, in the case of the person quoted above, or within the United States. And field marketing groups quite often have room for one or two entry-level people, who might start out doing logistical work—getting materials out to the field in a timely fashion, adapting written materials for non–U.S. audiences, assisting with setting up all sorts of promotions, chasing down suppliers to make sure they adhere to the required delivery deadlines, and so forth. Having good people skills is an absolute requirement. Technical knowledge is not. Being reliable, not letting things fall between the cracks, is very important. A tip: Sales managers absolutely want input into who their field marketing person is going to be. If you are an admin or a project coordinator or whatever, and a sales manager gets to know how reliable, energetic, dependable, and smart you are, you have a big leg up in getting a field marketing position because from the sales manager's point of view, these abilitiesare requirements. Smart sales managers and reps have a way of finding out who's on the short list to support them through field marketing, and since successful salespeople generally call the shots in nontechnical debates, they can wield a lot of positive influence on your behalf.

Marketing Communications—If you're a creative person and can write well, marketing communications may interest you. It's known in the industry as "marcom," and here's how the corporate director of marketing communications of a large high-tech hardware company defined it:

> It essentially supports marketing objectives and business objectives. It includes advertising, press relations, trade shows, literature, sales promotion. And I think that with the advent of new technology, there'll be all kinds of multimedia that will fall into the same category as well.

It's unusual for people in marcom to have technical degrees. Often they come from the fine arts or journalism, but many educational backgrounds are represented in this area. One of the most successful people I've met in the field only has a high school degree; she got started in high tech as an admin. Marcom people sometimes feel that the high-tech culture, being centered so much on technology, tends to downgrade their function. One marcom specialist said:

> I think that frankly we have a lot more marketing knowledge than some of the "pure" marketing people. For example, we often find them claiming, "Here is my marketing plan. We're going to go to these trade shows. We're going to do these advertisements." But I'll say, "That's not a marketing plan; those are marcom tactics." We are very much internal consultants to management right now. In our company, for example, few people understand the concept of branding. In a consumer company the whole purpose is to establish a brand and have brand loyalty. Most of our managers will not be able to tell you what a brand is or its importance. Our [marcom] people are strong functional experts with many years of experience.

The marketing communications function is a very important pillar of support for the effort to move products from the high-tech company to the customers. For example, high tech, with its rapidly changing and developing technology, depends to an unusual degree on the many trade shows during the course of a year. Just selecting which shows to be in and which ones to let go by is an important decision. Having a presence at a trade show costs money—in rental of space for a booth, construction or rental of the booth itself, staffing and transportation, and design and printing of special promotional literature, videos, and other media. Budgets are limited, so it's essential that, once the decision is made to participate, everything proceed smoothly. Organizing for such an event starts many months in advance, and a high-tech company is usually in different stages of planning for a number of events at one time. All this takes a high degree of intelligence and energy to bring off successfully, and the marcom function has the prime responsibility for making it happen. Of course, just showing your products at an event isn't the real objective: You want to be sure to identify prospects who will buy your product.

Operations

Operations is a general term for a manufacturing group or department that is responsible for supplier management, materials procurement, and distribution of the products. Thus it watches the supply line going into the manufacturing plant, and the distribution of the finished product to the shipping department. In some companies forecasting of orders and shipments is located in this department. Different enterprises may organize these functions in different ways, but they all exist somewhere in the high-tech company.

Supplier management means assuring that the suppliers of "raw materials" or, in the case of the high-tech industry, components, understand the requirements of the company as regards delivery times, cost, and specifications of the materials. Distribution involves the maintenance of sufficient stocks in a warehouse to meet forecasted needs, and the physical fulfillment of orders. Materials management, or procurement, involves taking the sales forecast, comparing it to existing stocks, and placing a purchase order with suppliers to make up the difference.

After a while materials managers get adept at reading forecasts and building in safety factors; there's nothing as agonizing as having a huge backlog of orders that can't be shipped because the materials to make the products weren't ordered in time. As the deputy operations manager of a software company told me:

> You have to use your judgment. Am I usually high? Low? Then, on top of that you have to concern yourself with pricing models, especially in the software print world. If you buy small quantities, you pay high prices; if you buy a lot, the prices drop way down. You have to balance the risk of obsolescence with the unit cost considerations.

There's a lot of "outsourcing" going on in high tech; forecasting, supplier management, and materials management will almost certainly remain inside the company because they are too sensitive and too critical to entrust to an outside contractor. As the deputy operations manager told me:

> We sat around a table and decided what we wanted to keep in-house. The actual picking up of boxes is what we're getting rid of. We'll tell the outside company how many boxes of [product] to stock, and then the order will come in. It will then be transferred electronically to the distribution company, and they will pull it off the shelf. When they run out of material, they'll call us up and tell us they need more.

The operations department of a high-tech company tends to be fairly stable, so new employment opportunities don't open up often in the more mature companies. Younger companies are a different matter, however. Since operations is not a glamorous field, a lot of the hiring is done through networking. If you learn of an opportunity in this field that is "entry level," or if you have some special experience that would qualify you for a midlevel position, you can try for it without fearing that the lack of a technical background will keep you out of the running. What is important for operations positions is familiarity with math, having an orderly mind, and being able to deal with multiple projects at once. The educational background of the deputy operations manager cited above was in literature and economics. He also has an MBA. One person reporting to him has an undergraduate degree in literature.

Typically operations is a tough area to get out of, so far as moving into sales or marketing. Product management and finance, however, are areas into which operations people have moved.

Order Operations

Order ops, as it is usually called, involves assuring that orders are "clean" when they are put into the system. *Clean* means that the part numbers match the written description of the product, that financial terms are met (letters of credit, net thirty days, and so forth), and that there's nothing about the order that would cause it to be held up. If the order is international, an order ops person will ensure that all export licenses are in order. Order ops people deal with the factory or operations groups to know when orders are likely to ship; they will inform sales reps if a delay is anticipated. They know how long, for example, it takes to build a certain type of computer; if there is a shortage looming, the order ops staff will try to set some priorities to ensure that the most time-sensitive orders are filled first. This often calls for reserves of patience and diplomacy.

Order ops is another function that is not glamorous, but it does have the advantage of being nontechnical and of affording those who work in it the chance to learn a lot about the innards of a fast-paced business. And you needn't have a technical degree or a business degree. At one of the hardware companies in Silicon Valley a young woman working in international order operations has a degree in art history—just like the woman I described meeting in the introduction to this book. Starting as a receptionist, after a year she became an admin, and a year after that entered order ops. She told me:

> When I was an admin I really didn't know anything about the products. Now I know all the desktops, desksides, peripherals, software. So I really think I've picked up a lot working here. And you have to know it in order to answer questions from customers, and sometimes from the salespeople as well.

From her present position, it would be possible to go in any of several different directions, including finance, operations, and even sales.

Press/Public Relations

Press relations is a very sensitive area, which is often, but not always, under the control of the marcom department. What is said about a high-tech company in the press can have an immediate effect on customers' decisions to buy or not, the company's stock price, and the morale of its employees. The marcom people handling PR must have good judgment, know how to deal with the media, and have fast access to all levels of management in the company. Major sales deals and other important company announcements are examples of issues handled by the press relations people.

The public relations function is closely allied to, and sometimes merged with, press relations. I spoke with a young woman about what was involved in her PR job:

> The world of PR is different within every organization. A lot of smaller companies have PR doing marketing and advertising. The bigger the company, the more you break it out, and the more focused the job is. PR is responsible for creating a positive image of the company that you're representing. And there's a secondary aspect of the job, which is to keep the public, the media, and the partners informed on all the products you're announcing. All the things you're doing within the company—strategic relations, crisis management. . . . Typically we find out about a crisis situation before it really happens. For example, some partner is going to do something that might affect your public image. So you work collaboratively with them in developing key messages and so on, and then you get it cleared at the highest level of the organization.

PR people work directly with technical people:

> We also work in product promotion; it's an important part of PR. We work directly with the product marketing people. We are in charge of the way the products are perceived. We write the release to position the product within the industry and get it out to key reporters. We want to show that the industry is behind it, so we try to leverage all our partners to support the announcement.
>
> There's a top tier of press and analysts that work with us

on a regular basis, that represent really important publications or organizations.

Working in PR gives you exposure to all levels in the company and is considered to be one of the fastest-moving, most demanding, and classiest activities in high tech:

> A lot of my friends are really jealous of my job because I get to do a lot of things right away that a lot of people take years to do, talking to the press and to the other media. Also my in-company points of contact are mainly at the director and the executive level of the organization because the communication needs to come from the top down. And you get to travel—for example, when there's a trade show. I've been to New York a few times, and also to Germany. We also deal a lot with the legal department, since we need to know what we can refer to publicly and what is off limits.

I asked another PR person about her background and what skills she thought were most important for her job:

> I was a communications major and had a business minor. For PR, the main thing is to be able to articulate your ideas. You need to have strong writing skills, strong interpersonal skills, and have strong powers of persuasion. You always get into discussions with reporters. Most corporate PR people start at an agency because that's where you get experience all across the board, from writing press releases to developing media plans—they have to do it all. Then typically if you've been doing that for a while and a company likes you, you can go in-house at that company.

Product literature isn't as sensitive as a press release, but in the high-tech industry it's an extremely important adjunct to the selling process. Brochures, catalogs, and "spec sheets"—glossy sheets of paper, often with color photographs or drawings, giving the basic specifications of the product—are very expensive to produce. Marcom has the task of working with the product managers and business managers to determine when a new product will "FCS," or "first customer ship;" the literature (or "collateral material," in the high-tech jargon) has to be planned and coordinated accordingly.

So regardless of what the techies may think, marketing communications is an important function and one that welcomes nontechnical people. If you are good at writing, can handle lots of things at once, are articulate, and can appear composed in the midst of chaos, this is a career area you should seriously consider.

Product Marketing

Product marketing is primarily a function that links the prospective users of a product to the designers of the product in such a fashion as to make the product more useful, efficient, and cost-effective. A person working in this field takes responsibility for seeing this is done for one or more products—sometimes a complete line of products, such as computer monitors or printers. This function ensures that the company does not end up designing and building products that are a delight to the engineers who designed them but that no one will want to buy. The product marketer will work closely with the design engineers on the one hand and with the potential customers on the other, while taking into consideration the special requirements of the production and service people. If everything works out, the final products will be useful, easy to build and service, priced competitively, and will sell well in the market.

Because they work so closely with highly technical engineers, most product marketing folks have acquired a good command of the technology. Companies like to see people who are engineers or who have MBAs, or ideally both, in such positions. But many do not have technical training, other than what they've picked up along the way. I can think offhand of three successful product marketing managers whose formal educations were in political science, biology, and history.

Product marketing, sometimes also called product management, is a very important job in the high-tech industry. It's usually the sort of position one works up to, though in a smaller company or for a product that management considers ancillary to the main product line you might be able to move into a product marketing position more easily. Product marketing managers own a particular product from its inception to its demise.

They are responsible for learning the market's requirements, overseeing all aspects of production and introduction of the product to the market, and forecasting initial and ongoing sales. They stick with the product until its EOL (end of life). It's sort of like being a geneticist, obstetrician, pediatrician, internist, gerontologist, and undertaker for the product, only these roles occur within a period of twelve months to a couple of years—a typical product lifetime—instead of the length of a human lifetime.

A senior product manager at a large software company commented about her job:

> Oftentimes I think the great secret is that common sense is the most important element. I've got this product idea, and I need to get it built. How do I do that? It's just a question of breaking the problem down into steps and being very organized on how to move it along the path. Sometimes you understand what the steps are, and sometimes you don't. Now obviously, over time, as you do this over and over again with different products, you start to get some expertise in the process.

So here's a job that is usually classified as being very technical, and it turns out to involve the application of common sense. Having common sense doesn't depend on having a technical degree, as we learn from another product manager:

> My senior year I wrote a significant paper on Jacques Maritain and French Christian existentialism in the twentieth century. Now that qualified me to cocktail waitress in just about any bar on this planet. I got into high tech through a combination of luck and a mentor—a woman who works for this company. I went to a temp agency and started temping, and one of the assignments I got was for a two-week stint to replace a receptionist for a technical support group. It lasted a lot longer than two weeks.
>
> I went in there my first day, having never touched a PC. I discovered there was a whole pile of letters that they had gotten that no one had ever had time to address. So, I will never forget this, my boss said, "Well okay, let's get you a word processor. . . ." She turned out to be a wonderful mentor.
>
> The first area of skill that I built was helping customers with nontechnical problems. I think maybe there were forty

people in the company then. We continued to grow and put in systems like bills of materials and those kinds of things. I did it all. I also got involved with shipping stuff internationally, so when the company determined that it really needed to be international, there was only one person in the company—me—who even had any inkling of a notion of the international market. They brought in a very high level director, a European woman, and she and I together decided that we would do a simultaneous launch of this product in twelve countries. She had the relationship with the distributors. But that relationship absolutely depended on us delivering the products, and so I sort of became a de facto product manager.

When the company was acquired, the buying company had no product managers. It was a pretty new concept at the time, and again it was a matter of luck that we merged with this company. The company made the decision that they wanted product managers, and I had as much of a skill set for that at that point as anybody. Plus, I had product knowledge because I had been working with the product for so long. So there I was.

Product marketing managers work closely with other kinds of marketing people as well as the sales force and engineers who actually design the product.

What happens to a product when there's inadequate product management? Ever heard of the Lisa? It was a computer that Apple designed for the office and enterprise market. It was a pretty good computer, too—but it never caught on. Apple never figured out how to position the Lisa in the office environment, and eventually Lisa was put to sleep. Or take the Mac. Apple was justifiably proud of the Mac. Apple was making scads of money on this line of products and didn't notice that the combination of Windows and Intel—Wintel, in industry parlance—was taking a commanding lead in the desktop market. It was at this point that Apple was approached by companies who wanted to make Mac clones. In retrospect, this would have been a sensible approach because the clones would have broadened the market, and the software developers would have kept writing applications to the Mac operating systems. But the product managers objected; clones would have forced Apple to lower its prices (inevitable, in a commodity market) to stay competitive with the clone market.

By the time Apple's management finally did decide to license clones the company was already on a steep slope and starting to slide. But the clone strategy worked. The market broadened, and the developers decided to stick around. Then Apple terminated its clone strategy and sales nosedived.

Where you have good product management, you see successes like Intuit's Quicken. Microsoft tried to buy Intuit; when this became impossible, it set out to defeat Intuit in the home financial management market. But Intuit not only survived, it grew and soundly outmarketed Microsoft. Not many companies can make that claim.

Or take Sun Microsystems's Jini™ software. This is a product that, simply put, enables a device with a chip and some software—phone, TV, computer, radio, toaster, printer, scanner—to hook on to the Internet, announce that it is there, and make use of any other device on the Internet as if that other device were directly linked to the first device. Your cell phone could conceivably make use of a supercomputer. Sun developed this rather startling technology a few years ago, but judged that the market wasn't ready for it. So the company kept it secret and continued development work. It was announced in 1998, when the convergence of computing and all sorts of communications, entertainment, and commercial technologies indicated that the time was ripe.

Dell Computer Corporation has had phenomenal success in selling its computers by phone and on the Internet. Why? Because the company was quick to perceive that, in the computer hardware industry, it is very difficult to establish a competitive advantage over other companies' products. To tackle this challenge, Dell decided to concentrate not only on making great products, but on making them customized and easy to buy through their direct sales model. This strategy cuts out all sorts of distribution overhead, resulting in greater profits, while offering customers the appealing option of a made-to-order computer.

All the above examples involved major product management decisions. Some worked, and some didn't.

Getting into this branch of marketing is like so much in the high-tech world: if you want the job, you can make it happen. To move into product marketing you've got to get into the industry and then get to know the products and the people who are doing the product marketing function. When you're ready, you sell yourself to the VP or director in charge—with several of your new product marketing friends as a cheering section.

One interesting thing product managers get to do is have direct contact with customers. This is perhaps their greatest utility to the high-tech company. The sales force is interested in selling; the product manager is (or should be) interested in selling something the customer needs, and he or she is in a position to do something if the customer looks confused or aghast when a new product concept is presented to him. This is not to malign the sales force; obviously it prefers to sell what the customer wants, but it must sell whatever is on the price list. The product marketing managers make sure the products on the price list make sense.

Project Coordinator

This is a really interesting position for low-tech people seeking to join the industry because it's a low- to mid-level position that is about as flexible in its content as possible. The content varies according to the needs of the department at any particular moment. A software company might decide to develop an online catalog of all its applications. A project team is formed, but the team really consists of the project coordinator, who knows what has to be done and establishes a schedule, creates a list of people he or she has to enlist in the effort, and then manages the total project. I know many people who got started in high tech because they found out through a friend that such-and-such a company was looking for a project coordinator. Because the job content is so amorphous, it's hard for a company to get very specific about the experience a candidate has to bring to the party, so it's a natural for smart, organized people who may lack high-tech experience but who have other relevant skills.

Here's the comment of a young woman who was originally hired as an admin, but who shortly thereafter switched to a project coordinator position:

> A project coordinator is someone who will be handed a project. In my case it was the [software product] solutions portfolio. We needed to produce a guide to third-party implementations for the product. The two or three most important skills that you have to bring to bear when you're trying to do a project like that are, first, teamwork, because you'll be working with a

number of people on a certain project. Organizational skills and communication skills are definitely a must. I think those are the three highest priorities that I see. It's very interesting because you learn on the job. That's definitely how I got to where I am today.

So being a project coordinator is not only a relatively easy gateway into high tech, it can also afford the opportunity for a lot of growth and learning.

Sales

You'll hear of two kinds of sales reps in high tech: outside reps and inside reps. Inside reps stay within the company building and work on the telephone; they don't go to the customers' sites. Generally they are more junior than the outside reps, who may have five years or more of sales experience when they start their jobs. You can read more about this position later on, under "Telesales." But outside or inside, it's absolutely amazing to discover where high-tech salespeople come from. They can have undergraduate majors in psychology, history, economics, literature, and many other nontechnical fields. There seems to be no logical connection whatsoever between the formal education these people have had and their present jobs. It's a field that lends itself naturally to people who are gregarious and confident. That being the case, there's no reason you too can't edge yourself into a sales position. It takes some time because you do need a fairly thorough understanding of your product to be successful, and you have to sell a sales manager on your ability to acquire such an understanding in order to get the job in the first place. But selling yourself as a sales type of person is far more important than having product knowledge, which can be acquired fairly easily. The two most important prerequisites for a high-tech sales job are an ability to communicate effectively with different sorts of people and a strong desire to make a lot of money. The least important requirement is having high-tech educational credentials.

A software salesman told me of his nontechnical background:

> I started in English and philosophy, and after a year of fifteenth century poets and even more abstract philosophy, I figured out

> that that was not what I wanted to do. So I moved into psychology and finished up with a B.S. in psychology. I got a graduate research assistant slot and started out for the Ph.D. But I realized after a year and a half I didn't want to do that either. I got my M.A. so I'd have something to show for my effort. And I also got an M.A. in broadcasting. It was easy to do because many of the credits were the same. I started in broadcasting with a student internship, but after three months they started paying me. And then I moved to a TV station and worked there for two and a half years.

Not terribly high-tech so far, is it? But he got interested in computers, took a course or two, and got into a PR job. Then he convinced a sales manager to take a chance on him. He ended up making over $150,000 a year in high-tech sales. He also won some special bonuses, and a trip to a fancy resort with his wife when he made his yearly sales quota. From sales he became the director of marketing for a high-tech company. I don't think it is doing the sales folks a discourtesy to observe that money is by far their prime motivator. For this reason, you will rarely find a salesperson willing to discuss trends in the industry, or anything else that might serve to take time from selling.

Another successful sales rep who spoke with me sells software that manages advertising traffic on Web sites, scheduling, delivering, and preparing reports for the advertisers on the success of their campaigns. He told me:

> I came to this job with a nontechnical background. So I had to acquire knowledge quickly. At an Internet start-up the ramp-up time is expected to be less than it would be at some of the big companies. With these start-ups there's no dedicated resource for training people, no dedicated time frame in which someone is allowed to learn the ropes. You have to use the resources available on an informal basis in order to acquire that knowledge. There are plenty of engineers in the company. You just build a rapport and take some of their time—ask them to explain a configuration, for example.

Being an inside sales rep is a good starting point for people who have energy and optimism. A young man doing inside sales for an Internet company said:

> There's a lot of prospecting involved, a lot of identifying qualified leads. In some companies the inside sales rep does that initial work and then works with the outside sales rep to close the deal. There are people who become outside sales reps in a year, but they won't make nearly the amount of money as an inside salesperson who has been at that same job for five years. So it just depends.
>
> I guess the most important skill at the entry level is persistence. That's going to be the number one character trait that you have to have to be successful. Number two is probably communication, and that's a high-tech skill in general. . . . The reality is that if you don't do well in communication you won't do well in high tech, period. So what is communication? It's, for example, when I call a customer and I find out they've increased or decreased their purchasing budget. There may be five other people in our organization who have to know. That may be the VP of sales, the director of sales, finance, all kinds of people who may do projection of our revenues. So you are responsible for getting the information to them quickly, and clearly, so they can act on it.

High-tech salespeople who are good—and there is no other kind because the bad ones leave quickly—are the princes and princesses of the industry. A youthful hardware salesman who might be in his late twenties had no doubts about that:

> Let me tell you something. I'm making my numbers and exceeding them every quarter. No one—*no one*—can get between me and my customers. I can tell the sales manager to [expletive deleted] and he'll do it because I'm making him successful. Not even the president of this company can get in my way. If he does, I'm out of here like a flash.

Yes, there's a little bit of ego there all right. Yet deservedly so. Salespeople are out on the thin edge. If it weren't for them, the business would stop. The relationships they have with their customers are their precious stock in trade, and there are moments when salespeople feel that no one at home is supporting them, that they are carrying the weight of the entire enterprise on their shoulders. Prices are revised and no one tells them. Products are EOL'd (EOL stands for "end of life"; when a product is

EOL'd, it is withdrawn from the price list and is no longer available for sale) and no one tells them.

The outside sales representative is seldom found in the office; sales reps are usually out calling on customers, who might be either end users of the products or one of the sales channels. When a sales rep is in the office, he or she may be following up to be sure that orders put into the system have been filled and shipped. Generally, sales reps are not credited with a sale until the product has actually been shipped to the customer, and toward the end of the quarter (high-tech salespeople have quarterly goals) the salesperson may spend a good deal of time in the office on the phone, trying to iron out any obstacles that may delay or prevent shipment. These can include credit issues, competition from other salespeople for a limited supply of a product, or simply a misplaced or lost order.

I asked a woman who sells overseas for the U.S. subsidiary of a Japanese high-tech company (and who had no high-tech background at all before she got the job, though she speaks fluent Spanish) to comment on the positives and negatives of high-tech sales as a career:

> I'd say the good points are, when you are selling, everybody loves you. The company loves you, everything is great, and that certainly is positive. You feel like you get instant gratification when you get a new contract, when you get a new program. It's very gratifying to have people compliment you and say that you're doing a great job. And another good part of sales is the contact with people.
>
> Sometimes, though, I get a little tired of the emphasis on the numbers. Because the bottom line is always numbers, at least with the management. . . .

Sales reps are expected to be very familiar with the products they sell, but this does not mean a deep technical knowledge of how they are constructed. They read up on new products, consult with product managers to find out what the new products will do and when the expected "FCS" or "first customer shipment" date will be. High-tech sales is a tense, exciting (sometimes nerve-wracking) and very well paid occupation. When you live like that you love the people that support you and you hate the people that get in your way. Which brings us to another category of jobs.

Sales Support

The sales support function has many faces in the high-tech industry, but the idea is the same—make the system work so that the sales reps can do their job and make their numbers, so everybody can get paid next week. It's an important position because it takes care of everything, with respect to the customer, that the salesperson does not have the time to do. And that's a lot. Salespeople, after all, live or die by their numbers, which are calculated every quarter by merciless number crunchers located in a basement office. If the salesperson misses his or her number by a small amount, maybe the quota was set a bit too high, or maybe the customer decided he would defer half his order until the next quarter, but it's a sure thing then. If the salesperson misses the number by a very large amount, perhaps this was due to some uncontrollable act of God—a seismic event, perhaps, or a collapse of the Asian market (as happened to a sales manager friend of mine in 1998). Or perhaps the customer's treasurer ran off with the bank account, the customer is bankrupt, and thus a major sale on which the rep had been working for months simply fell through.

No matter. The sales manager develops a steely look in his eye. The number crunchers send their statements upstairs. The moving finger writes, and having writ, moves on. *Mene, mene, tekel, upharsin.* Thou hast been weighed in the balances and art found wanting. One quarter is excusable. Two quarters aren't. (Although in the case of my sales manager friend, they were—perhaps because he never whined to his boss. Sales VPs admire stoicism.)

The only sure friend the sales rep has is his or her sales support person. When the rep is off chasing more business, the sales support person may be finding out what happened to an order the customer swore was faxed in last Tuesday. Or maybe the support person is alerting the rep to a problem with the credit people or the lawyers. The sales support person is the person the customers can call when the rep is on the road, who'll find the answer and get things done. The sales support people may visit customers with the sales rep from time to time, or they may work entirely behind the scenes. They may host customer visits. They can ensure that the customer got the literature, the price book, the promotional stuff that the marketing people developed. These people get as close to the customer as one can get without being in sales. They don't make as much money as a sales rep, but they're reasonably well paid. And they don't have a sales quota to make every quarter, which means they sleep well the nights preceding the solstices and equinoxes. It's nice work.

Strategic Alliance Manager

I asked a very successful woman who heads a strategic sales alliance group in a hardware company to comment on what kind of background made for success in the job, and what the job involved:

> What we develop in this company really is only a piece of the complete solution the customer wants to buy. So if we want to go into a company and tell them about our products, that's important, of course, but at the end of the day what the customer cares about is solving a business issue. And the way they do that is to have a complete portfolio of services and products. So we've recognized that if we partner with other companies—the other developers and manufacturers—and go in jointly, we're more likely to win the business. These partnerships are called strategic alliances, and they're really important to success.
>
> As for what the job requires, you have to be able to learn how to learn. You have to have an analytical view, and an ability to look at data and understand how to interpret it. It's also important to understand how to ask questions, and how to listen. These basic skills are important. What I learned in college [psychology and mathematics] hasn't helped me much. Well, it got me jobs, but the subjects I studied really have no relevance.

One of the main advantages is, of course, that the company uses its partners as its eyes and ears to find out about opportunities for future business. As the manager told me:

> Of course, all the other [companies] are doing the same kinds of things. But what you find out is that the quality of the relationship does influence what they do. Take [company] for example. They will always say that they want a completely level playing field, that they are not going to prefer one partner to another.

And that's what everyone will say. But the reality is that people do business with people they like. So we all get invited to the party. The differentiating factor is who you want to do business with. My staff does a couple of things. One is, they build the contacts between our company and the ISVs. The other is to build the relationship . . . well, it's really a virtual team, one that involves people around the world. So we can share successes and best practices. And the third thing they do is working with end-user accounts.

A lot of this is international, and one of the things we've discovered is that different cultures need communication in different ways. . . . Part of what we do is modeling behavior. For example, we'll go to Argentina and actually show them how to go about winning a deal like this. Being there and doing it, doing it with them. Writing it down, sending it out in e-mail, put it on a company Web site. We also hold team meetings where we bring the worldwide team together and get them to make a presentation on their successes.

As for the skills I look for [in new hires] . . . the communications stuff is really important. Probably the most important are skills of persuasion. Such skills are essential for working within this company and with outside partners, people who don't report to us directly. To help them understand why it makes sense for them . . . in a sense, to toe the line.

As is apparent, it's not a requirement to have a high-tech degree to be in partnership management, but a few years of experience in marketing or sales are usually required before you can be seriously considered a contender for such a position.

Technical Editor

Because the writer is a team member, he or she may acquire a biased view of the importance of the project. That is when the technical editor becomes useful. This position is available to writers who love and have the patience for editing. The technical editor not only performs normal editing functions, but also can provide perspective, consistency, and continuity. Individual project teams are formed and dissolve as their missions are accomplished or, sometimes, abandoned. The editor can assure that every bit of information is incorporated into the final document in a manner that is both internally coherent and that matches with what the company has said elsewhere. As an editor said to me:

> As the editor for the software development framework project, when I get a submission (of material) I go through and make sure that what is said is consistent with, and not out of proportion to, things we've said about similar topics elsewhere. And I get edited as well; we have an outside contractor that looks at what I do. In an important project like this there are a number of levels [of control]. We issue a document of about five hundred pages every six months. If anything is out of sync, it can cause major problems.

An ex-teacher I met who became a technical writer, and then an editor, has an undergraduate degree in English and a master's in teaching, both from the University of California at Berkeley. This woman went on to describe the change from the public school environment in these words:

> I taught junior high and high school for eleven years. At the end of those eleven years, education was starting its grand decline in California. Money was drying up. Class sizes were getting larger. I was thirty-five years old. I had applied for other positions in the school district that would have been promotions for me, and had not gotten them. I just felt that my options were really narrow.
> I got to feeling kind of depressed, actually. So I got a summer job at [company] doing technical editing and writing. I had never done that before. I met one of the engineers that worked there on a Sierra Club hike, and he gave me an introduction to some of the people there.
> I was very relaxed! I didn't have all the stresses and strains of classroom teaching, like grading papers, and teaching writing. On the job I could get everything done really fast. All the people in my department were amazed. And I was amazed that they were amazed.

Technical Writing

The high-tech industry desperately needs qualified writers to prepare all sorts of technical manuals and publications. Though there has been some improvement in the quality of technical writing in recent years, as good writers have become aware of the need, the quality of much of the (very extensive) literature in the industry is still poor. Some colleges and junior colleges offer courses and even degrees in technical writing. Since customers rely on online manuals to get their systems to work, many companies are pushing to improve the quality of their manuals. Documentation is one of the most critical products a high-tech company provides; in fact, documentation not only helps the user understand and work with the product, it also serves the high-tech manufacturer as marketing and product positioning material. So the writing positions are of great importance to the sales organization in a high-tech company, and technical writers are treated with respect and are well paid.

The technical writer usually works as a full member of a team that is developing a new product (or some part of one) or working to improve an existing product. He or she gets the raw information directly from the technical, service, and marketing people on the team. Writers tend to become quite expert in the areas of technology they work on. Most do not come from a technical background, but rather have degrees in literature or some other area of the humanities.

Telesales

"Telesales? Ugh!" I heard you say it. But as you may have gathered from the section on sales, telesales is very different in high tech than in other businesses. Instead of pushing cosmetics or exercise equipment on people who would rather enjoy their dinner than talk to you, working in telesales in a high-tech company involves informing potential customers of products or services that can save their companies money, by making their businesses more efficient. The initial reaction of the people you call is to listen to you, not to hang up. This in itself is a wonderful thing, but better still is that, after a while, you will have mastered every aspect of the product you're selling, through fielding the questions people throw at you. And you start to learn about your customers' businesses as well—their names, what they do, where they are located, and so forth. In other words, you start to get immersed in the high-tech world outside your company through the end users of whatever it is you're selling. If you're calling the MIS (management information systems) manager at a large financial institution, you will start to learn about that field and feel comfortable dealing with it. If you're dealing with the medical industry, you start to get immersed in its particular concerns. More important still, you have been on the front line dealing with customers and can add this experience to your resume. This is the sort of thing that greatly impresses hiring managers. A young telesales group manager offered this view of the job, and of how he came into the high-tech industry:

> I had no connection with technology, and when I graduated from college I worked for a high-tech ad agency because I thought I wanted to get into the advertising world. [I] spent a year and a half in the market research department. . . . That was sort of my connection with technology. Then I went to the clients, and I actually worked for a distributor selling products. I was selling mass storage devices. I learned about them just from reading.
>
> I had good communication skills and good people skills, and I understood how to sell ideas and things. One of the big disconnects in the high-tech industry is that most marketing folks and most salespeople in the early days percolated up from engineering. And companies found that those didn't always make the best marketing and sales folks because you can teach people technology, but you can't teach people to handle rejection. In assessing candidates, I think you look for attitude and you look for understanding the sales process and you look for people that understand what it takes to be successful. You can teach the technology.
>
> To be successful it takes a couple of things. I think you need to be really self-motivated. You encounter a lot of rejection in sales. It's compounded when you're on the phone. . . .
>
> Someone that had a really high aptitude for sales, yet had no technical background—I would want to hire that person. The ideal candidate has experience, is successful, and has a technical background, but let's face it—that doesn't always happen.

Please do not make the mistake of passing over an entry-level position in telesales. It is absolutely possible to use such a job to acquire a solid base of knowledge about a company, its products, the market, and lots of other useful business knowledge. It is a good stepping stone to an outside sales position or a job in marketing.

Training Positions

Many nontechnical people, particularly ex-teachers, get their start in the high-tech industry through the training function. Some companies offer lists of courses that would put a small college to shame. These courses are aimed at employees of the company; they can cover everything from how to put on an effective presentation to negotiating with the Japanese to time management to computer systems administration. Other companies offer some in-house courses and contract outside for others. Some of these courses provide skills needed to do your job better, and some are highly technical, aimed at providing specific knowledge in a new technical area.

All this training has to be planned, developed, and delivered. Nontechnical people coming from the world of education may feel more comfortable in this environment than in high-tech areas, at least initially. A woman who teaches at a large corporation said:

> I had a bachelor's degree in social services. Then I went into retailing for a year and a half before going to a telecommunications company as a specialist hooking up phone lines. I was there for about six and a half years. There was no formal training there, just an older person I worked with. But I kept getting promoted, from specialist to senior specialist to supervisor. Finally I ended up as a manager. But I didn't get technical because I was more on the people side. The company had no money for training, so they asked for volunteers to train others. My department was going along well, so I volunteered. The training was in how to supervise. I went from that job into high tech, as a training specialist.
>
> Training is a big field now in high tech, and it's more secure than it was ten years ago.

This emphasis on short-term, highly specific training reflects a belief on the part of the top management of the company that colleges and universities simply cannot keep up with the rapid pace of technological change. In the words of a senior manager, "Technical education has become vocational education. Would you go to school for four years to learn how to turn out cranks for automobiles?" If you're going to be negotiating with the Japanese next month, the company wants you to get the basic dos and don'ts in a couple of days, not a couple of semesters.

There are several nice things about being in the training field in high tech: People generally show up to class on time; they don't throw spitballs; and they are motivated. And you don't have to deal with the board of education or angry parents. Also the pay tends to be good. People who have labored in the public education field for years appreciate these differences. As one person told me:

> I worked twelve years for the [state] educational system. When I came out here I decided I wanted to make the move from the public sector to the private sector in spades. My perception was that the public sector is kind of looked down on, like a forty-seventh cousin. People who work in it are not generally regarded as professionals.
>
> I had a horrible feeling that I wasn't qualified for anything. People talked to me about going into training because I had an education background. I ended up going into a software development training program. I didn't teach in it; I managed the program. I remember the first visit I took with my boss to meet with a bunch of engineers. He told me not to talk to anybody. He didn't want them to know I didn't know anything.

Web Designer

High-tech companies are generally linked to the Internet and also have a private company intranet. Each of these networks will usually have many Web sites, through which the public (via the Internet) and employees (via the intranet) can have access to all sorts of information. Both networks are of great—even critical—importance to high-tech companies, causing an amazing growth in the number of jobs available for creative, nontechnical people. The need for home page designs; written, graphic, and musical content; and constant maintenance of the content has built a huge demand for artists and writers, but almost any smart and creative person can position himself or herself for a Web-related job. As a writer working in an Internet software start-up told me:

> Basically there's such a demand for people now that they are looking for people with sociology degrees, English degrees,

philosophy degrees—people who have degrees where their minds are trained to think. The Internet is so hot in so many companies and has so many applications that it doesn't matter what your degree is. That kind of gives you motivation to find out about the Internet, and having learned about it could be your key to getting into a company.

The man who founded the multimedia department of a prominent university had some interesting comments about career opportunities in this field:

I could share from my experience of teaching an introduction to multimedia class for about three years. Most of the people that come there are people who see some possibility and come to find out about it. They don't know much about what they're getting into. And what they see is a place where they can take their passion: it might be writing; it might be photography; it might be dance; it might be film. Most of them were resigned to giving up those interests because they got a job in banking or what have you.

They discover that they can integrate their skills from banking, or financial services, or medical services, or whatever with their appreciation of the aesthetic, and have a job where they can make money. New jobs are occurring, but it's not all work for technical people. It's a whole range of work that's enabled by technology. They come to find out about the technology, so they can go and integrate their other interests into a career, and it happens.

Here's how one young woman started her high-tech career via the Web:

One of the things I do is manage an internal Web site covering almost everything except engineering, for the employees in the division. One way to start out in this field is as an intern. You take a couple of classes in graphic design and learn how to use Photoshop. And learn how to use HTML. You've got to have a good eye for how information is organized. That's what it really boils down to. Being a temp has been very successful for many people. A couple of my friends have gone from waiting tables to learning the basics of some office productivity packages out there, when they did temp work.

> Don't be afraid of a computer. It really is not all that scary. If you learn how to think logically, you'll be fine. There are plenty of books out there like *HTML for Dummies* or *Writing Cool HTML Documents*. You can look information up on the Web—there's lots and lots of resources there—and you can ask anyone [in the company]. Ninety-five percent of everything I know has come on the job.

The number of jobs that involve the Internet is constantly growing. One expert said:

> The demand always exceeds supply. I just did a presentation in San Francisco; the mayor held a multimedia summit to talk about the future of the multimedia industry in San Francisco. I interviewed a bunch of CEOs, mostly from multimedia companies in San Francisco, to find out what they're doing, why they're in San Francisco. Everyone's biggest challenge is finding enough qualified people for the work we have. . . . There's a huge demand, and it's growing. The growth figures for San Francisco alone are enormous. Something like 35,000 jobs in the city of San Francisco.
>
> How much time does it take for someone to get up to speed for a multimedia job? It depends on what people are bringing to the job. I've seen it happen in six months, and I've seen it take two years for people to land a job.

Not only does the Internet provide a lot of opportunities today, the growth rate is extraordinary. A report from the U.S. Department of Commerce reported in early 1998 that Internet traffic was doubling every 100 days. Contributing to this is the fact that the Internet is a global phenomenon. The expert quoted above went on to observe:

> It's great for people who have international experience. Because one of the challenges that everybody faces as they do business in the world of the Internet is how to make things appropriate for multiple cultures. The big centers of growth for the next five years are not in the U.S. They are all around. . . . Like Brazil is on the verge of being this enormous new market. They are privatizing their telecommunications industry, which will be the largest privatization of any industry in history.

There's a lot of opportunity for people who are willing to think creatively, put things together, and take some risks to get something that satisfies them, that nourishes them, that allows them to have a life they want. I think that the more of us that pursue that goal, the more possible it will be for others, and I'm really heartened by that. It's one of the things that really turns me on about the whole deal.

STARTING YOUR HIGH-TECH CAREER

Getting a job in high tech is almost always the result of a process that you initiate and execute, and over which you have control. Sometimes you hear of people who simply lucked into a position. That doesn't happen all that frequently, though if you'll recall chapter 1, that's how Jenny moved from being a waitress into high tech. In part, getting into high tech is a numbers game: the more possibilities you uncover, the greater the chances you'll find your first high-tech job. In large part, too, it's a connections game—a game of networking, if you prefer that term. Some people consider *networking* to be an overworked term, but it's a very powerful tool.

There's also an element of magic in getting that first job. You've got to work a little legerdemain to present yourself in the very best light for a particular opportunity. For a manager, the hiring process is an interruption of his or her normal activities—one of the reasons managers hate employee turnover is that they have to devote unscheduled time to getting replacements on board—so you, as a candidate, can help make the hiring manager's life a bit easier by making the process as easy and natural as possible. Show them what they've been looking for, so they can get back to their work.

Effective communication is an essential element as well. We low-tech people have the raw material to enable us to be good communicators. In the hiring interviews we must demonstrate not only how smart, energetic, and articulate we are, but also that we can listen thoughtfully to others, ask sensible questions, and generally build empathy with the interviewers.

Finally, the job hunt involves quite a bit of methodical work—to uncover those opportunities, to learn about the histories, products, technologies, finances, employees, and cultures of the companies you target. You *can* get a job in the high-tech industry if you go about it with intelligence, determination, and persistence. As a Shakespeare scholar working in a software company told me:

> Persistence is what counts. If you want to do something, never give up under any circumstances! If you want to do it, it will happen. Most people give up remarkably easily. They have too much of their own ego in it.

So, as you can see, there's a lot to accomplishing that first step towards your high-tech career. Keep this bit of wisdom in mind, though: your task is to work your way into high tech with any job you can get—that is, any job with a decent group of people, in a company that'll survive for at least a year. So you should pursue a broad range of opportunities without becoming so attached to any single one that you neglect the overall effort. Once you are inside the walls of high tech, you begin a magical transformation. After three to six months, you will start to understand and use the jargon. You'll have a handle on the job, and your confidence will rise. In that short time, you'll be able to network and build up a fair number of contacts within your company and in other companies as well. You'll feel more comfortable with the pace, have had some encouragement from your manager, and you'll have taken some courses. And you're now in quite a different position from when you were outside the walls wondering how on earth you could ever get in.

People seeking to switch into high tech from another type of work find it difficult to accept this notion of "taking anything," especially if that means taking a pay cut. But if you stick it out for a year, you'll be ready, armed, and able to maneuver relatively easily within the world of high tech. In short, you can get a better job, at higher pay. Once inside, from any position at all, you can maneuver into the position you really want and deserve. A channels-marketing person reported, "My cousin got in by being a janitor. Then he moved on from there. Get in any way you can; there's no one way to do it."

But we're getting a bit ahead of ourselves here—first you have to secure a job offer, and that involves carefully planning out a job-search strategy. There are seven basic stages of the high-tech job search for people with no technical background:

1. Appraising yourself
2. Adjusting your attitude
3. Acquiring background experience
4. Networking and identifying opportunities
5. Compiling a resume and cover letter
6. Handling interviews
7. Coming to closure

Before moving on to a discussion of each of these stages, there's one piece of advice I'd like to equip you with. During the course of your job search, you'll accumulate many important documents, including any personality, interest, or skill assessments that you have completed, records of your networking contacts, and copies of each resume you've used, with a notation of the recipients. It's essential that you organize this material so that, when the need arises (and believe me—it will), you can quickly put your hand on any given document or phone number. I'd also strongly recommend keeping an appointment calendar so that you don't forget to place important phone calls; this calendar will end up being your running record of the campaign. And when in doubt, write it down.

APPRAISING YOURSELF

Your personal plan starts with your making as objective an appraisal as possible of the kind of person you are. For some people, high tech may not be a good choice. It's better to find this out before you start the job-hunt process. Several tests, properly administered and interpreted, can help you learn a lot about yourself and how you'd fit into the environment of the high tech industry. These tests are administered by employment and career counselors and psychologists, and are worth investigating. The four that are usually at the top of any career counselor's list are these:

- **California Psychological Inventory (CPI).** This well-known test was developed in the early 1960s. It predicts performance in social settings, school, and job functions. The test gives eighteen different scales on which behavior is measured, and it is easy to relate these to work life. The score for a particular scale might indicate that the person taking the test likes to work in settings where the job is tightly defined and where there are clear milestones for measuring progress. This is usually not the case in the high-tech industry.

- **Campbell Interest and Skill Survey (CISS).** This test measures both your interest in various occupations and your self-reported skills. This inclusion of the skills component is one of the strengths of this test; the subject might have strong interest in being a world-class photographer, but the interest alone will not carry him very far if he lacks some of the skills required by this occupation. On the other hand, the test can indicate occupational areas where he may have both a high level of interest and high skill levels. These might offer much more rewarding careers.

- **The Sixteen Personality Factor Questionnaire (16PF).** This test was constructed by examining most of the adjectives by which personal behavior is defined and reducing them to a set of sixteen bipolar characteristics that underlie them. Some of these are Reserved/Outgoing, Trusting/Suspicious, Practical/Imaginative, and Group-Dependent/Self-Sufficient.

- **The Myers-Briggs Type Indicator.** This is a very well-known self-administered test of personality and preferences. It has been used for many years in career counseling. It asks a series of questions about your preferences, forcing you to choose between two answers. The person interpreting the results will describe where your scores place you in four areas: Introversion-Extroversion, Intuitive-Sensing, Thinking-Feeling, and Perceiving-Judging.

It is possible to overdo this analytical stuff, in my opinion. It should be emphasized that those deeply involved in research into the roles and interplay of vocational interests, abilities, and personality characteristics as predictors of happiness and success in a career admit that much work remains to be done in the field of career assessment. (An interesting book on the topic of career assessment is *The Clinical Practice of Career Assessment*, by Rodney L. Lowman.) I think that, besides taking the tests mentioned above, you should solicit your friends' opinions of how happy you would be in an industry where the work is characterized by the following:

- little direction from superiors
- long hours

- work judged by peers
- high pay
- fast pace
- responsibility for own career development
- speaking before groups
- pushiness valued
- multicultural, multiethnic
- performance valued rather than style
- a predominantly young staff
- too little time to do the work
- frequent retraining/reeducation necessary
- must deal with technical people
- frequent reorganization
- use of jargon
- dealing with uncertainty

These are not trivial issues. Career counselors, using the tests mentioned above, can also help you with these issues. Take the fast pace and the need for frequent retraining, for example. Things change so quickly in high tech that at times it may seem unwise to take a day or two for a training class, out of fear that you'll miss something, like an important meeting, for example. It's not unheard of for high-tech employees to put off attending a class several times. They want to take it, and they know they should take it, but they feel too busy, too rushed, too harassed.

People do grow and adjust to new environments, so I don't think that you'll necessarily reach the conclusion that you *shouldn't* proceed with your high-tech job search, at least until you see firsthand what is available for you out there. But you ought to start out with your eyes open.

ADJUSTING YOUR ATTITUDE

The transition to a high-tech career can be relatively easy, or extremely demanding. If you're young, don't have a mortgage or family responsibilities, don't have much of a track record and thus aren't branded as "an insurance guy" or "a schoolteacher," chances are you can get a low-level job in high tech fairly easily. But what if you've been already employed for several years? What if you've been working in outside sales for a lumber company for fifteen years, or as an insurance agent or small retailer, and want to switch into high tech? Can it be done?

Sure, it can be done—but there's a price. The effort you'll have to make will be greater; the initial sacrifice of income may be greater (though probably not for the teacher). You'll have to build up some computer-related knowledge (and this may be difficult, given other demands on your time), and the cultural changes will be more wrenching. You're going to find that people will ask you why you want to give up a perfectly decent living as an insurance agent to go into customer service. This is a legitimate question, and you should have an answer before you launch yourself into the job-hunting process. The more familiar you are with what the new job entails and how it fits into your transferable skills, the less difficulty you'll have giving a plausible answer. An approach that goes down well with high-tech hiring managers is that you were feeling bored at your present job, you analyzed your skills and compared them to what was needed in the new position, and discovered that there was a really good fit.

You might have to adjust your attitude at the very start of your job search. A job hunt can be frustrating. It might help at the beginning to reframe your perspective and say that it's really a process of making friends with lots of interesting people and developing more knowledge about the high-tech industry. Here and there along the way, job possibilities start to appear, and sooner or later along comes the one you land. This has happened to many people I've met, who are now working as professionals in the high-tech industry. These people came from the following fields and educational backgrounds, among others: teaching (elementary, junior high, high school, college, music, art), educational administration, television and newspaper journalism, urban planning, psychology, secretarial, theology, machine tool operation, and sports broadcasting. They wanted to get in and they made it happen. A young woman told me:

> Initially I was turned down for the position. There were two hundred people applying for two positions. I didn't get it the first round. But then the husband of the person who got the job was transferred, so I followed up and told them I was still interested. They called me back a month later for a second

round [of interviews]. It was difficult because there were a lot of qualified people applying for it. For example, one of the people had six years of experience in the telecommunications field. I had zero experience. I think mostly they hired me because they liked me.

There's that persistence again. That may be the most important arrow in your job-hunting quiver. Expect to make moving into the high-tech industry a real campaign, one that will involve a lot of learning and growing. You may luck out and get a wonderful job in a week, but it's more likely that it will take three to six months to get inside the walls of the industry. Do you have the persistence to go the distance and not give up?

The industry uses the term *mindset* a lot. It means having the right attitude and locking it in the way a navigator sets a course with a radio compass on an airplane. The plane flies through storm clouds and high winds, but it's locked on course and it gets to its destination. You've got to do the same thing.

ACQUIRING BACKGROUND EXPERIENCE

What are hiring managers really looking for? Peter Vozas, a senior recruiting manager, was quoted in an article in the *San Jose Mercury News* as saying that "the first criterion is, Can the person fill the position without a lot of training?" Many people seeking to join the high-tech industry have expected that they could show up with the raw material (their intelligence, personality, and attitude) and count on some sort of formal training to help them get up to speed in the high-tech job. That just doesn't happen. The training possibilities for career advancement are many, but there's nothing formal for the beginner to get started in that first job. It's not that you can't get in without being 100 percent qualified for the tasks, but the training you get will be all "OJT"—on-the-job training.

So how do you bridge the experience gap when you're trying to get that first job? If you know very little or nothing about computers, you should spend some time and effort to get up to date. These days you won't be at all credible to a hiring manager unless you feel at home with some hardware and software terms, understand what networking (in the computer, not the job-hunting sense) is all about, and can surf the Internet.

Many courses are available that can give you the basic information you'll need to feel comfortable with the technology, so you'll come across to hiring managers as being confident that you can do the job. Don't try to anticipate the kind of work you'll end up doing in your high-tech job and pursue specific courses that seem to strengthen your background. Instead, look for courses that will make you familiar in a general way with technology. They're given everywhere: at adult education centers, colleges and junior colleges, career centers, and so forth. If you're a creative sort, pay particular attention to courses dealing with the development of multimedia for the Internet. You'll find that the staff of such institutions will be very willing to discuss the sorts of courses that can help you beef up your resume. Simply looking at a catalog that lists such courses can be very frustrating because there are so many and the descriptions won't mean much to you at this stage. A half hour spent with a teacher is time well spent. (By the way, teachers of such courses are often very well connected to the industry and can serve as good leads for networking contacts.)

Of course, there are specific skills that would prove helpful. It's useful to be able to work with spreadsheets, so a short course on Excel might be indicated. You've got to know how to use a word processing program, such as Microsoft Word or WordPerfect, as well. And if you're onto what looks like a really good prospect and you find out from an insider that the company is really hot on some particular application, there's nothing that prevents you from burning some midnight oil to get at least some familiarity with it. That way you can claim a degree of familiarity and keep a straight face. I knew someone who did this with a great application called FileMaker Pro. She got access to this software and spent a few days working through the book. This application was widely used in her target company, and she had it on her resume.

I asked a young friend who is in Internet marketing what advice he would give to someone seeking to build up some high-tech credentials. He said:

> I would say the first thing you could do is subscribe to some magazine, . . . *Computerworld*, for example, and *Red Herring*, which looks at technology in general from a smaller-company focus. Then they might get onto the Internet and subscribe to

things like Infoseek, or Yahoo!, or whatever. Track technology. My Yahoo! has a technology section. Someone who wants to get into high tech must be comfortable using a Web browser and getting on the Internet. And then build your network. Start with your friends; rely on them to give you the names of people. Then if they say such and such a software company is an interesting company, you might say, "Gee, I don't know anything about that, but I'll go look it up." You do a Web search and go to their home page and find out what they do.

If you don't already own a computer, get one—they're cheap and getting cheaper every week. Many people love the Mac and have used one in school. I love the Mac too, but my advice is to buy a PC—that is, a "Wintel" machine (Windows 98 or NT operating system with an Intel chip). Apple has become a niche player in the industry, and despite recent successes, it does not appear that the company will be a major factor in the market. You can also buy (or be given) a used computer, but if you do, be sure it has full documentation, ideally in book form (i.e., not on your hard drive). You should read the documentation and try to understand it. Get a computer glossary—an excellent one is *Computer Glossary: The Complete Illustrated Dictionary*, by Allan Freedman. This glossary is really a minicourse on computers, software, and the evolution of the industry, and you'll find yourself referring to it frequently.

You can attend some meetings of a local user group—people who meet perhaps once a month, or more frequently, brought together by a common interest in a type of computer or software. Newspapers generally have announcements of such meetings in their "calendar of events" sections. User groups are often top-heavy with techies, but they are usually very approachable folks and are willing to coach and answer questions.

If you don't have much confidence in your ability to participate in a hiring interview, consider taking a one- or two-day course in interviewing skills. It is absolutely essential that you be able to come across in an interview as organized, calm, collected, thoughtful, having a sense of humor, and being able to listen. Many people have good backgrounds but shoot themselves in the foot during the interview process. One common fault is that interviewees want to show they know everything about the new job— that they've done their homework, and that the new assignment would hold no mystery for them. They come across as too eager to please, and often they don't let the interviewer get a word in edgewise. Being able to present ideas effectively is one of the most important skill requirements of nontechnical persons in the high-tech industry; the hiring interview is the first opportunity you'll have to do this in any particular company. But being able to listen attentively is also of critical importance during the interview. If possible, try to find a course that videotapes you in an interview situation so that you can observe your unconscious mannerisms, body language, how well you maintain eye contact, and so forth. Ironically, even being overconfident can be detrimental. A friend of mine has tried for a long time to switch companies to get a director's position. I happened to know the last hiring manager he interviewed with, and asked him why my friend didn't get the job. His reply: "He just had an answer for everything. He's smart, all right, but no one knows everything."

Go for some informational interviews early on. If you already have some general background in computers and software and you know what kind of job you're after, you can use information gleaned from your early interviews to zero in on specific courses—ones that would boost your chances of getting it. This is really not such a difficult task, assuming that you're currently employed (or otherwise have no problem paying the rent) and are not under time pressure. Once you've had a few informational interviews with people in local companies, you'll have a better idea of what background you'll need. For example, many high-tech companies furnish each of their employees with a desktop computer, usually networked. And these are usually not PCs, but something more powerful, such as a workstation. You'll learn that it will be helpful if you can present yourself in job interviews as being familiar not only with PCs, but also with workstations, their user interfaces, and (for example) their graphics capabilities. Of course, if you're interviewing with a systems manufacturer, such as Silicon Graphics or Sun Microsystems, it's best if you can say that you've worked with the particular company's own machines, and sometimes this is a specific requirement. But it's not too hard to go from being familiar with one type of workstation to another, and what many people have done is to get familiar with company A's machines in the few days between the time they find out about the job opportunity and the time they apply for the job. They do this sometimes through the contacts they have developed within company A itself through networking, sometimes through user groups. Five hours with a particular machine can give you the confidence to look a hiring manager in the eye and say, "Yes, I've used your systems."

To summarize, if you're not at all familiar with computer technology:

- Take a couple of courses
- Get a computer with documentation and learn how it works
- Get a glossary and learn the terms relating to your computer
- Join a user group
- Acquire interviewing skills
- Go for some early informational interviews

Getting some high-tech background requires you to spend a few dollars and invest some time. The payoff is that you'll have fun, forget that you were ever afraid of technology, and you'll increase your chances of landing a great job.

NETWORKING AND IDENTIFYING OPPORTUNITIES

There are four possible ways of identifying opportunities in high tech: cold approaches, responding to advertisements in the newspaper or on the Internet, temping or contracting, and hearing about the opportunity from a friend or acquaintance (networking). We'll talk about networking first, although it may not be possible as your first approach, because it can be useful in the other approaches.

Networking

Of all the techniques used to get a first job in the high-tech industry, networking is by far the most effective. It's effective for your second and third jobs as well. In fact, networking is a technique that should be practiced throughout your career in high tech. Effective networking involves contacting an insider—someone who knows the company, can give you inside information about the job, about what the hiring manager is like, what the salary level is likely to be, who the competition is, and lots of other useful information. Plus, a really good networking contact can boost your chances by enthusiastically recommending you for the job. This is probably as good a place as any to let you know that many high-tech companies pay their employees $1000 or more for successful referrals, once your insider gets your resume into the system and you get the job.

A newspaper article in the *San Jose Mercury News* about a company in Silicon Valley that had grown just past the start-up phase, but that wasn't large enough to attract talent just by virtue of its reputation, revealed that over one third of all new hires came from employee referrals. These insider referrals were the single largest source of job candidates.

A network is for life—not just for the first job. You've got to keep it nourished and growing. A senior vice president of a Fortune 200 high-tech company told me:

> If I look at my own career, luck of course comes in from time to time, obviously, but apart from that my network has helped me the most, the relationships I built along the way. It's very true in our industry. There's an HP network, a Prime network, a DEC network, and I guess there's an IBM network. So, very often when it comes down to filling jobs, people are asked if they know anybody who can do the job. Our president, that's the way he fills the positions. He goes to his board members and his staff and asks them if they know anyone they've worked with in the past who could do this job. More jobs come this way than from recruitment agencies. After all, these agencies themselves rely on the same networks.
>
> So it's all about who you know. And the network tends to be trusted because if you're referred, then you're a known quantity. The person who refers you knows that his reputation is on the line. My own key moves were influenced by the contacts I had. Positions were suggested to me that I might be interested in. You've got to build associations with your peers. The loners very often can go so far and they fizzle and fail. It may not be right, it may not be fair in some cases, but when the going gets tough, the network will keep an eye out for you. At this point, networks are essential to a career.

Going the classified ad or Web listing route, while sometimes necessary, is fraught with difficulties. You can only control the process to a limited degree and it's highly competitive. There has to be a better way, and there is. Networking is the way to go. Networking, properly executed, is the Ponzi financial pyramid scheme redirected to the job market, with you as the exclusive beneficiary. The networking process opens doors that remain closed to others. It can get you in front of a hiring manager *after* you have researched the job through conversations with your future colleagues—a situation that works to your advantage since it arms you with vital information about the job and the tough questions you might encounter in the interview before you talk to the hiring manager.

How do you start your network? Starting with someone in the high-tech industry helps, but isn't necessary, because everyone knows someone who knows someone who works in high tech. Try starting with friends or family members. A man who is with a software company that develops Internet applications managed at age thirty to make a switch from a career in food export to a high-tech inside sales job by networking.

> My wife and I bought a house on the Peninsula and I was commuting to Oakland and thinking, you know, if we live in Silicon Valley, I might as well give this high tech a try. And through a network of friends I learned about [company]. A friend of a friend actually working at [company] at the time recommended me to the VP of sales, who interviewed me and gave me the job.

A young woman got a lead through a family member.

> I went to a family-type function and . . . I talked to my cousin who was working at [company]. She had been there four years. And I said I was interested in marketing and she said she would look around. She got me an informational interview first, and they happened to be hiring, so it kind of worked out that way.

A woman in her mid-forties who had just come into high tech told me:

> You cannot underestimate the value of contacts and friends that you have. It was a key issue for me, and it was a key issue for someone who moved out of my group about six months ago. . . . She moved right on to another group easily because

> she had made so many great contacts with people, and she had some opportunities pop up for her because she networks really well, and I don't network that well. . . . I would advise people to somehow make those connections. They could be through, I don't know, clubs and groups and friends you already have. Not putting pressure on them, not even necessarily giving them a resume, but just putting the word out that you're interested.

Having an insider contact makes you credible to the hiring manager, for the very simple reason that some of your contact's credibility rubs off on you. A woman who is now a VP of a software company spent some years working in state government as a human resources consultant:

> Then I decided I really wanted to get into the private sector, which is very difficult to do. But because I had been consulting I had some clients in the private sector with whom I had some credibility. After a while, one of them asked me to come in. I was lucky. It's awfully tough. The private sector is really skeptical of the public sector.

How do you start your network if you've just arrived in town and really don't know anyone? I suggest that you join a health club and get to know the members. You can be sure that some of them come from the high-tech industry—or know someone who does. If you can't afford the fees, just ask if you can put a notice up on the members' board with those little tear-off phone numbers on it. Explain in the notice that you need ten minutes of someone's time for an informational interview. All you need is one reply. (Your social life may pick up as well.)

Professional and alumni associations are great places to start as well, as are outdoors organizations such as bicycling and hiking clubs. Professors who teach computer courses will be delighted to give you names, especially if you have been thoughtful enough to have enrolled in their courses before asking. They are just as interested in networking as you are, since it is likely that much of their next European vacation will be funded through consulting to high-tech companies. And don't overlook the sorts of groups that read aloud from the Great Books. There's a Shakespeare group in Silicon Valley totally composed of high-tech worshipers of the Bard. Dental hygienists can be great sources of names.

You get the picture: You can start a network anywhere. The keys to successful networking are to be organized and to follow through. If you get two names from one of your contacts (and don't leave the premises until you do), write them down in your job-hunt records and call those people as soon as you can. From each person you talk to, you should get two or three additional names of people in the high-tech industry, or in professions that deal with high-tech companies. If you are disciplined and work at it, you'll acquire an extensive and unique list of contacts.

A friend of mine who had been working in the public sector for a few years decided to seek greener pastures. Here's how she started out:

> I went to the Career Resource Center in Palo Alto [now located in Cupertino, California]. As part of the membership, I received a three-hour appointment with a career counselor. So I made my appointment and I walked in and she had out her magic markers and pens to do tests, and I told her to put it away, I had an agenda. So she said, OK, what is it? I said, First of all I want three names from whom I can get informational interviews so that I can get started. Second, I want to know what to wear for the interviews in Silicon Valley. So we spent some time talking about that. I got my three names and she told me to dress like Barbara Walters.
>
> I wanted some names. I didn't want to have to cold call. It scared me to death. So I had three names and I could start. I damn near got a job offer on the first one. And everywhere I went I got three more names. It's a pyramid. So you can continue meeting people. That's the only way I could think of to get a job.
>
> I had a whole spiral notebook, and I kept the names and the notes on the interviews, directions on how to get there, what happened, when I'd written the thank-you notes—all kinds of things.

The whole aim of networking is to start out gleaning information, collecting names, and learning who's who and eventually to end up with a job offer; or, more likely, to find out from one of your contacts that someone in his division expects to have a job opening in the next few weeks, either because that function is expanding or because someone is leaving the company or otherwise moving on. This is precious information because you are (you hope) the only person in the whole world outside the company who knows of this. When that happy moment arrives, there's one thing you must keep in mind: Don't stop the networking process. Even if the position seems tailor-made for you, remember: there is no "ideal job." Any job that gets you in is good. Keep plugging along, investigating all the other strands of your network, rigorously taking names and making notes. As mentioned elsewhere in this chapter, the high-tech hiring process is often chaotic, and strange things can happen. The hiring manager can get sick and die. He can also quit the company the day before he was to mail the offer to you (this is less forgivable than dying). Some Wall Street analyst has a bad day at the track and the next day puts out a "hold" recommendation on the company's stock, which plummets, and a hiring freeze is announced. A rival manager gets his hands on the head count. It goes on and on, and it has all happened before. So be optimistic but keep your network alive and humming until the moment you get an offer letter.

When you get the offer letter, by the way, countersign it immediately, make a copy, take the original to the post office, and mail it return receipt requested. That way you've got a proven hiring date. High-tech jobs have been known to vanish after the offer letter went out, and managers have been known to deny receiving an applicant's acceptance letter. Not that this would ever be the case with *your* boss.

Cold Approaches

In a cold approach, you get a list of companies and broadcast your resume as though you were an ancient Sumerian sowing grain. Of course, you don't just mindlessly mail your resume. You get together a list of companies that have needs you think you can fill, get the names of the presidents, and send each one a letter with a resume. The theory is that the president would pass your letter and resume down to the hiring manager, who would pay more attention to your resume because it came from above, not through the U.S. Postal Service or via e-mail.

For most people this shotgun approach doesn't work. After all, the president's secretary might be just as likely to send the letter to the HR department, where it might be treated to a ritual burning. But if the letter and resume are really professionally turned out and the company really needs someone with your background, it might work. It did happen to a man I know. He was a consumer-products marketing person from Chicago who was trying to get a marketing slot in a high-tech company in California. He simply sent his resume and cover letter directly to the president of the company. The president instructed his marketing VP to arrange interviews for the candidate. I don't know how that story came out, but I hope the marketing VP didn't feel offended that the candidate was cleverly trying to go outside the hiring process. Usually line managers aren't bothered by such things.

Here's a cold approach story that definitely ended in success, the experience of a woman who is an international marketing manager for a well-known software company. She is European, has an undergraduate degree from a European university, and started her job search in the second year of her MBA program at a large Midwestern state university:

> I started with Standard and Poor's directory. There were a couple of drawbacks [to my job search]. One was that I was trying to get a job in California from the Midwest. The second was that, as a foreigner, I didn't have a work visa. So I decided I could only get results if I put a lot of work into it. I did about two to three months' research at the library on companies, and I gathered all together maybe 500 names and addresses of these companies. Then I narrowed it down to about 270, using the product, the size of the company, their management. For example, if they had women in management. Also their location and whether they had international operations.
>
> I sent a mailing that included a cover letter and a resume to all these companies. I said that I would be in California on a certain week and that I'd be available for an interview. Then I did a telephone follow-up to approximately 150 of those. And I did get twenty interviews for the week.
>
> After these interviews I followed up and got some second interviews. And by spring break I had another set of companies that was much more narrow, maybe around seventy, where I had followed up with letters or sometimes phone calls. I got around fifteen interviews at spring break.
>
> Out of these companies two or three were interested. One of them flew me back. I did get two job offers. So by mid-April I did have a job. It was a small company, and I was the first marketing person on board.

If you have the time and are a methodical person, you can conduct such a campaign and maybe you'll be successful. You be the judge. The chances are that you would not thereby limit your ability to use other approaches to the same companies. The flood of applicants is great, and HR memories are short. You would not be jeopardizing your chances for normal approaches to the company later on. In fact, most recruiters and HR folks would admire the determination of someone who makes every effort to get into the company—if they have useful qualifications. On the other hand, there are a few real losers—pests—out there who just don't understand "no." I know that you are not a loser, so don't worry about being considered a pest.

Job Listings on the Internet

Thanks to the spread of the Internet and the fact that every company worth working for has a Web site with a jobs listing on it, it's gotten a lot easier to find out where the opportunities are and what the companies offering them are all about. Unlike the situation with newspaper ads, there's not a lot of incremental cost in listing all the job opportunities the company has available on the Web site, from ones demanding highly specific knowledge, skills, and experience, to the entry-level positions. Many companies keep their Web sites up-to-date with general information about the companies' businesses, their top officials, history, and so forth. The Web is a great resource.

That doesn't mean, however, that the Web is going to replace networking and having an inside connection. Seeing a job posting on the Web that appeals to you can be terrifically exciting; the Internet itself lends an air of immediacy that very understandably affects the job seeker. That's especially true if these magic words are appended to the job title: "Available for immediate hiring." I wish I had a dollar for every time I've heard from young people, "I sent my resume in. It's a perfect match, and they said they wanted to fill it immediately. And it's been five weeks. . . ." If everyone who applied for a job through the Web listing also had an insider who could let him or her know the real scoop on what's happening, there would be a lot less disappointment and a lot less emotional investment in any one particular listing.

Job postings on the Web require you to send your resume not to the hiring manager but to a recruiter. What happens from that point on can be a subject of great mystery and sorrow. The job posting gives you some of the information you need about the position, but by no means all the information. The recruiter has spoken with the hiring manager and has pretty strong feelings about the type of person that would succeed in the job. The recruiter's reading of your resume, and his or her gut feeling about you, will determine whether you get called in for an initial interview. Since lots of resumes are usually submitted for a particular position, even though you may feel that your qualifications suit you very well for doing the work, several other candidates may—by luck, perhaps, or perhaps though having some internal pull—have edged you out.

Newspaper Ads

Strangely enough, it's often rather difficult to find out exactly what the requirements are for jobs in the high-tech industry. This is especially true if you found out about the job through a newspaper ad. Not only may the language used to describe the position be unfamiliar to you, but because these ads are expensive, not much detail is included about the work itself. Of course, if it's the sort of job you've already done, you'll recognize the word cues and feel comfortable about responding. On the other hand, there is liable to be too much detail of the kind that leads you to exclude yourself from consideration. That's why looking at ads, or Web listings, for that matter, is usually a depressing experience. You may think you've found something that fits, and then bop! There's the kicker. "Five years' experience" or "MBA preferred."

One thing to remember about this sort of hurdle is that the HR people sometimes encourage the hiring manager to insert these phrases in an attempt to build in some sort of prescreening, not because the requirement is really important for the job. Thus, if you can present yourself in a powerful and intriguing way, you can often circumvent these "requirements." You must be able to show the hiring manager that you appreciate the complexity and importance of the position, that you clearly bring to the table those fundamental skills required to do the job well, and that you are quietly confident of your ability to grow rapidly in the job and take care of any ancillary requirements. Gaining the trust of the hiring manager is worth virtually everything. If you meet the basic skill level of the position and the hiring manager thinks you can do the job and likes you, you have won.

But in many cases—especially for jobs advertised in the very expensive newspaper "display" ads used by the larger employers—the job requires a very technical person; it's not a case of HR putting hurdles in your way. You'll recognize such cases because they will always require a type of college or graduate degree you haven't got.

When looking for a high-tech job through classified ads, therefore, it can be much more rewarding to look for ads placed by companies that provide services to the larger, better known high-tech companies or that serve as sales channels for high-tech products. Don't ignore these ads because they are small. Such companies may be more impressed with your background than their larger, more blasé siblings in the high-tech world. And if you get a job, you will be accumulating valuable experience for your next move—be that within the same firm or to another company. A woman who is now in sales support in a large software company got her start in just this way:

> I saw a tiny ad in the paper for an electronic-component sales company, and that's how I got my first sales job. It was also my first introduction to anything high tech. Why was I hired? I like to think of myself as being intelligent and articulate and very upbeat, and they must have recognized in me the capacity to learn.

And learn she did, thereby positioning herself for a move to her present job:

> The vendors would come in on a weekly basis and do training, and I called vendors on a regular basis and asked questions. I also took a course in basic electronics to become fluent in the terminology.

Between the very large companies, where it's hard (but not impossible) to get a start, and the smaller sales and service firms, there is a very large number of midsize companies in the hardware, software, component, and networking businesses. Usually the strong growth is to be found in the healthy software and networking companies. Profit margins on computers and their components are growing slimmer every year, which is why these companies are seeking to reduce their costs, including payroll, any way they can. So check out the software industry.

You should always look at the ads to see which companies are hiring, and you should scour the smaller ads just as carefully as the larger ones. Sometimes you do find something interesting, and it looks as if there might be a fit. How do you verify this? Well, you can't, if it's a blind ad. Even when the company is identified, there's often a line in the ad that says "No Phone Calls." A phone call might be just what you need to verify the fit. But they said not to call! What should you do? This might be one of those cases when you should just do what you're not supposed to do. Once, a few years ago, my family held a garage sale. We had an ad in the local paper that gave the hours of the sale. I think we planned to open up around eight in the morning. By seven-thirty most of the good stuff was gone; the antique dealers had come and picked up the old rattan furniture and discolored mirrors, and a few ordinary citizens, veterans of the garage sale routine, had taken the kids' old desks and dressers. They didn't obey the rules we had set, so they beat out everybody else. Did we turn them away because it was unfair to take their money at seven-thirty rather than eight o'clock? Would you pass up a bag of money on the street? A com-pany will break its own rules, if breaking them will help the company, so a little aggressive marketing of yourself may well be in order.

If the company is in your vicinity, you can always drive over and try to charm the receptionist into giving you the name of someone to whom you can speak, working in the division where the job opportunity is located. That'll sometimes work, especially in smaller companies, and especially if you can claim with a straight face that you've driven an hour and that you really feel you've got the right stuff for the job but don't want to bother them with a resume unless there really is a pretty good fit. This is a serious suggestion, why not visit your prospective employer? At least you can learn the layout of the lobby. Maybe you'll arrange to get there at lunch time, and the receptionist can call someone over as the crowd leaves for the local deli. You might even get a free sandwich out of it. You can't tell what will happen, and occasionally what happens is nice.

And since a little pushiness is often valued in high tech, there is rarely a backlash to this kind of bold approach.

If you can't get the name of someone in the department you're interested in, you can focus your energies on finding someone who works in *any* department in the company who might serve as an insider for you. This is sort of a backwards networking. You've found the opportunity (you think), but you need to get more information, and you need to get known within the walls, even if you can't definitively breach them this early in the game. Unfortunately, you don't have a lot of time. The ad has been in the paper and hundreds of letters and resumes are accumulating in a pile somewhere in the HR department. These days, hundreds of resumes really do result from a single ad, particularly if the job is of a nontechnical nature. What you're really trying to do by hanging around the company that placed the ad is to make yourself stand out in any way short of setting fire to the place. If you can get inside, you might be able to accomplish this. If you can't get inside, then your best strategy is to research the company as thoroughly as you can in a couple of days and then write a "killer" letter (see pages 110–120 for a discussion of this topic).

Temping

If you remember your medieval French history (and who doesn't), the supposedly impregnable fortress of Mont-Saint-Michel on the Normandy coast was taken when a traitor let down a wicker basket from the ramparts one night and hauled up a bunch of English soldiers. They, in turn, opened the gates and let the English army in. Well, the temp agency can accomplish, and rather easily, the task of getting you up the ramparts and inside the walls of high tech. These outfits have greatly expanded their high-tech operations in recent years in response to the downsizing of the large computer companies, which has (of course) resulted in everyone else having to work much harder. The temp agencies can provide companies with a person for periods from a week to several months. Sometimes the company asks the temp agency to state explicitly in its job listings that the temp position may lead to a permanent one.

Temping provides a means for experienced workers who have been laid off to get an income stream, but it also provides an excellent channel for nontechnical people to find a first job in the high-tech industry. Of course, you may be thinking, these are only temporary jobs, and what you want is a permanent job. Well, stop. What you want is a bunch of things, including income, some benefits, a good opportunity to learn, and most importantly, a chance to network within the walls of high tech. Getting a high-tech job through a temp agency has become much easier in the boom period of the late nineties, but even during slack times temping can be a very effective arrangement. Some company somewhere will always be growing and need people. During the high-tech recession in the early nineties, a manager of a growing software company in the Northeast told me, "We use a fair amount of contract and temp help because we can't hire fast enough."

Then why shouldn't you just apply for the full-time job, if they're hiring so fast? You can and should do that, but you should also go through the temp agency because there is little or no competition for temp positions. That's right—temp positions tend to be "first come, first served." If you show up, you get the slot (assuming you have the minimum qualifications). No waiting, no muss, no fuss. Menial task? If the company is a good one, so what? Get in, do the menial work, and the fact that you started as a temp will make a great story when *Fortune* magazine does an article on you some day. Elsewhere in this book I told the story of a young philosophy grad who got a temp job at a start-up "for a week," doing filing. He's still at the same company, and he's got to be pulling down at least $175,000.

I spoke with a young woman who had been working for about two years with a start-up division within a Fortune 200 high-tech company. She started off her career by temping and then wisely started networking once inside the high-tech company:

> My undergraduate degree was in political science. I thought I wanted to be a paralegal, so I did that in New York for a while. Then I moved out here. I didn't want to be a paralegal, so I found a temp job as an administrator. I went to the temp agency and said, "I just need a job right now." And they said, "We have a position for you in the legal department of [company]." I had a six-month assignment initially—filing, copying, mailing, basically administrative work. But, as it turned out, I only stayed a month and a half.
>
> Through networking at the company, I was brought over

as a temp to join this business in a division where people were processing contracts and dealing with customers. Then I was given the task of writing summaries of all our contracts, and I also took responsibility for maintaining the price book.

When I left New York, people thought I was crazy. They said, "You have a great apartment. You have a great job. Why are you moving?" I've really bettered myself by taking a few steps down and then a few steps up. You really have to be confident and a little aggressive, and once the doors are open, find a fit for yourself. Once you get in the door, you can move around.

Temp agencies often serve the high-tech industry through separate divisions. The names may vary, but usually they amount to a technical division and an office services division. When you first approach a temp agency as a nontechnical person looking for a professional (nonclerical) position, you may find that the interviewer for the agency will not be sure which camp to put you in. He or she (usually a she) will want to avoid putting you in with the technical people, but may also feel that you don't belong with the office workers either. There are a few ways of handling this situation:

- **Go to more than one agency.** In fact, call all the agencies in your area—they're in the Yellow Pages—and register with each one that claims to serve the high-tech industry. Temp agencies are like real estate offices: they are nationally advertised but locally controlled, and your reception and how you are treated will depend a lot on the individual with whom you deal and the particular relationship the office has with the high-tech companies.

- **Be open to technical-sounding jobs the agency finds for you.** Often there will be entry-level jobs dealing with customers on the phone. If the company is sound and has a decent reputation (and you can discuss this frankly with the agency representative), don't hesitate to go after this kind of job. There's nothing like it for rapidly building knowledge and credibility. Remember, what you need are high-tech credentials, and manning the phones is an excellent way of getting them.

- **Be open to (some) office jobs.** Your decision should depend on whether the company is expanding, standing still, or contracting. You only want to get involved with a company with good prospects for survival and growth. If the temp agency can get you an interview as an admin in a software company that is growing like Topsy, and you like the people you'd be supporting, then it's well worth taking. If the admin position is at an older company, I'd say forget it—unless you like being an admin. If the position is a filing job, but it's for a start-up company and you like the people, take it. If the start-up is successful, you may be able to retire young. Even if the company grows only moderately, the chances are you will be able to expand into a better job within a relatively short period of time.

The temp agency rarely, if ever, really understands the full requirements for the job it has recruited you for. It has some kind of written description, but this may be even skimpier than the kind of job description you see posted on the Web. This can work to your advantage. When you show up on the doorstep of the hiring company at 9 A.M. Monday morning, the high-tech manager has no way of knowing anything much about you until you start to perform. The temp agency will have faxed over a resume, or maybe only called the hiring manager on the phone. She or he will not be as uptight about matching what you bring to the job with the qualifications of the ideal candidate. After all, it's "only a temp slot," right? The hiring manager's initial impression of you is based on your personality, how you're dressed, your energy, and other things you have direct control over. I cannot emphasize this enough: As a temp, the spirit and energy with which you carry out your tasks in the first few days will define you in the company's eyes. Even if you may not fully fit the company's requirements, the manager won't find that out until a week or more has gone by; by then, they're used to seeing you, and you may even have volunteered to do extra work after hours because of your rapidly growing interest in the business. Give the impression that you want to learn, to participate in the excitement, to grow as a (you hope) future member of the team. It's almost irresistible to a hard-worked manager to have someone that motivated. And if this approach doesn't work the first time, it will the second.

Contracting

The essential difference between the temp agency and the contracting agency is that, when working for a temp agency, you become an employee of the temp agency, which usually gives you some benefits. When you are a contractor, the contracting agency usually just lines up the deal for a fee. Payment, though, is sometimes made through the agency. Though this is changing, temping traditionally tends to place people into positions that are at the lower end of the income scale, such as administrators and other office help and folks that work on the loading dock. Contractors tend to have some sort of specific skill that the company needs for the moment, say, for the duration of a particular project.

In the bad old days of downsizing, some companies thought that they could dump lots of individuals, and then take them back on as contractors, thereby saving the 25 percent or so of the total compensation package in benefits they wouldn't have to pay for. But there are substantial legal pitfalls that lie in wait for such firms. If a contract employee is in all respects treated as if he or she were a full-time employee, except being denied the company's normal benefits package, the company might be forced (through a lawsuit) to make the contractor a true employee. As a result the employers these days have to be most careful about how they hire contractors; it has to be a real arm's-length deal, or it can end up costing the company a bundle. (A bit of trivia: This is the reason most companies don't give business cards to contractors. It make them look too much like "real" employees.) This can make it more difficult for you to go from a contract position to being a full-time employee, but if you do get such a position, it may enhance your chances of being taken on as a regular employee in another company—because you will have gained the experience you need. And there is always the chance that you may be able to come on board with your contracting company. That, of course, involves getting on really good terms with your boss and co-workers, and ideally becoming Mr. or Ms. Indispensable. That's what happened to a woman who recently turned a contracting position in customer service into a full-time job. She said:

> Actually, the original position was for a contractor. They brought me in to interview for the position, and it was telephone support. It wasn't exactly what I wanted to do, but I was willing to take it because I wanted to get into the company. In a couple of months they decided to make the job a real one. That's how I got to be a full employee.

A young woman I know came on board a high-tech company as a contractor. She had a great disposition and soon had built up a large circle of friendly colleagues. Because of the benefits, she desperately wanted to become a full-time employee. Her contract was initially for three months and then was extended for another three months. She invested some time in finding not one, but two marketing reqs (requisitioners) and in getting an interview with the hiring managers. But because of a reorganization within the company, no decision was made on filling either position. Then toward the end of her six-month contracting period, the word came down from on high: save money by terminating all contractors immediately. Fortunately, by now her reputation for efficient, cheerful, and useful work was so well established that her manager convinced a VP to let her at least finish the six months. He agreed, and she got hired into one of the full-time slots exactly three days before the contract was up.

One of the things this story illustrates is how easy it is for a company to terminate a contractor position. There's no risk of lawsuits, no payouts of any kind, because the agreement is between the company and the contracting agency, not with the person doing the work. Another thing the story illustrates is how it is possible to get into a permanent position through faith, persistence, and a bunch of colleagues who'll tell the hiring manager, "She's great!"

One advantage of contracting is that it may offer flexible hours. A young man who was a music major in college wanted to continue with his singing and didn't want to work "normal" high-tech business hours. He said:

> I don't want to be a contractor forever, but it's difficult to say when I will want to change. I enjoy this sort of flexibility because it lets me be here, have a decent job and make a respectable living, but it also gives me the freedom to pursue things on my own time. . . . I can make my own hours, and I enjoy the people I work with. There's another side, of course. Because I'm a contractor, my time isn't necessarily my own. I do projects for other people—the work comes in a flurry, or in loads. So I have an intense period, and then dry periods. It's not a stable sort of work scene. It's the lows that bother me. I'd rather be busy than bored.

Contracting one's services can lead to surprising changes in one's fortunes. For example, I know a man who was laid off from the HR department of a large Silicon Valley company a few years ago. His field was organizational development. The company still needed the work to be done after he was off the payroll, so they brought him back as a consultant for several months at a fairly high hourly fee. He discovered he liked being a consultant—especially when he networked his way into other assignments with other companies. His wife, who was initially terrified that her husband was on his own, caught the entrepreneurial fever, quit her job, and joined him. This two-person team has thrived doing contract work for companies and city governments for several years.

There are quite a few such companies that got started during the high-tech downsizing that took place in the beginning of the 1990s. Such companies themselves may be sources for employment in areas such as course development, training, communications skills, and sales. Other areas where new companies are being created to serve the high-tech industry include professional and technical employment recruiting. For some people the way into high tech starts by joining a temp or contracting agency as a recruiter, then acquiring additional credentials through a local college, and finally moving into a professional job such as HR specialist in a growing high-tech company.

Job Fairs

Occasionally, you may come across an announcement in the paper for a high-tech job fair. These ads imply that job seekers should arrive with resume in hand, ready to pass it along to company recruiters vying for the chance to sign you up. Unfortunately, these fairs rarely work this way (at least for nontechnical people). The recruiters are there for one reason only—to fill openings for highly technical people. For low-tech people, it's not only a waste of time, but can also be a degrading experience. You'll wait in lines, get shoved about and, when it is discovered that you're not an engineer, you'll be abruptly dismissed. Do yourself a favor and spend your research and networking time more wisely.

Before wrapping up this discussion of identifying opportunities, there's one bit of wisdom I'd like to share with you: There is no such thing as the ideal job. Quite often I've heard men and women enthuse about a position they just found out about at some company or other. They've either heard about the job from a friend, or they saw it posted on the Web. It's the ideal job because it seems exactly to fit their skills. The posting said that a knowledge of French was desirable, and they wrote their senior paper on Molière, or the job involves some (but not too much) international travel, the pay is excellent, and someone on the inside said there was no internal competition.

What often happens next is that person abandons the job-seeking *process* and focuses on pursuing the dream job. They invest so much emotion that there's nothing left for mundane activities such as reshaping the resume, delving into other possibilities, writing thank-you notes to members of their network, and so forth.

Very rarely does a person ever land the presumed ideal job—something always comes up—and all that emotion, energy, and hope is drained away, with nothing left but regret. Unfortunately, this often initiates negative thoughts, and the job-seeking process enters a downward spiral. What sorts of things can destroy the dream? Well, for one thing, there can be a hiring freeze—they happen from time to time. The "dream job" is there, but no one's allowed to fill it. Don't wait for the hiring freeze to thaw. During the wait managers can leave, priorities can change, the job can be "releveled" (that is, made a higher or lower level job, so that you are now either under- or overqualified for it), or the job can just disappear: "Yeah, we were gonna have a channels manager for France, but sales are down, and the finance guys say no way."

On those rare occasions when you actually land your dream job, you may find out it is too good to be true. That happened to a friend of mine very recently; she felt miserable and unappreciated in her job with a large software company because she didn't get a hoped-for promotion. A European company was looking for a business development manager who could speak German, and she was fluent in that language. There was a salary 40 percent over what she'd made in her existing job, the promise of European travel, and the possibility of a bonus. My friend had been looking around at other possibilities, but the moment the dream job came along she leaped at it and got it. She celebrated by buying a new car. After three months, however, the bloom had vanished from the rose. She'd lured a lot of important customers to the new company, which was busy striking deals, but these deals involved making commitments that the company had no means or intention of fulfilling. My friend started searching for another job. She would have resigned immediately, but she had the new car payments to consider. So there she was, stuck in her "dream job."

COMPILING A RESUME AND COVER LETTER

A resume is the one permanent document that precedes your visit to the company. It is a marketing document, designed to facilitate your movement toward interviews. A young woman in a hardware company told me:

> By marketing yourself I mean taking time to do a resume that looks good; having a personal appearance and demeanor that makes sense in that environment; and then just showing some tact in how you communicate with the employer. Attending to these finer points will help you sell yourself.

But a very senior manager has a different point of view about resumes:

> Q: How important is a resume?
>
> A: I don't consider it very important.
>
> Q: Why not?
>
> A: The information should be written there so that you're emphasizing your strengths, that's true. It is an information document. But in today's world the resume is usually not read by anybody. It goes into the HR department and it gets scanned into a database, on which key-word searches are done. It's a standard, canned program.
> And also this whole game of writing the resume and putting it in the right way, with all the right words, is BS. Because most people don't want to tell somebody they don't have a skill . . . and it's usually [because of the lack of] all those social skills of communication that a person doesn't get the job. So people lean on their resumes and then they don't get the job, and it's

supposedly because of something that's not on their resume. Hiring managers rarely tell them the truth: It's because you can't communicate very well. It's because I didn't like you when you came in the room. And you spilled coffee on my desk and you didn't apologize. It's that social, animalistic reaction that we have to another human being that decides whether we hire him or not. But the resume will be used as an excuse.

Q: Does this system (HR scanning of resumes) work pretty well in your company?

A: No (laughs). One of the reasons is that in this corporation the reality is that each HR group doesn't communicate with the others very well. There's not the self-discipline to create a pool of information that can be shared across the company. The system ought to work well, but it doesn't. So it depends very much on whom you know, your references, being a good guy—all of that stuff—word of mouth, being a friend of a friend when people are looking for somebody. Being somebody who can play with the team—that's the fundamental thing.

So what *is* the importance of the resume? A good one can't get you the job—only the hiring interview can do that—but a bad one will kill your chances of getting to the interview. If you must go through the HR department to get to the hiring manager, a good resume is critical. If the right buzzwords aren't on it, if your background doesn't seem to fit 100 percent, you're out before you get up to bat. A friend of mine who is a Hollywood actress told me once what it is like to go to an audition when there are a couple hundred other hungry aspirants for the assignment. "The casting company has decided it wants a woman between thirty-five and thirty-seven years old," she said, "five feet seven tall, and with just such a shade of hair, cut in just such a way. If you're thirty-four or thirty-eight, or if you're five feet seven and a half, or your hair is just a shade lighter or darker, they won't let you read. And this is for a spaghetti ad!"

That's what it's like sending a resume through the HR folks when you apply for a specific position. They've told you (through an advertisement) something about what they are looking for, but they haven't told you everything. The resume you submit is your educated guess, and it's a long shot. If you must go through HR, then you owe it to yourself to learn as much as possible about every requirement HR is looking for. You can only do this through talking to an insider, as outlined earlier in this chapter. Once you're armed with inside information, you can produce a resume that has a much better chance of getting past the initial screening. A good rule of thumb is that for every advertised position not requiring a technical degree, there will be two hundred external applicants and that HR will select five for passing through to the hiring manager. That's a $2^1/_2$ percent likelihood of success. Inside information ought to improve the odds in your favor to 10 percent or better.

If you can bypass HR and get directly to the hiring manager, the resume is less important than the fact that you were recommended by someone the hiring manager knows and trusts. But it still has *some* importance. For one thing, if the hiring manager likes you, guess what? He's going to turn your resume over to the HR people to include in their stack of likely candidates. HR people don't like to be bypassed and are not above sniping at you through alleged gaps in your background or other "weaknesses." This isn't likely to be fatal, however. It's the hiring manager who's going to have to live with you, not the HR people. The worst they can do is to point out that Candidate Doe seems to be slightly better qualified for the job than you are. Fortunately, by the time they've done this, you've already made your second or third visit to the office to meet and be interviewed by others—your potential future colleagues.

Your resume is important because it will be read by all of these people. In preparing it you can consult many of the standard guides, but in any case you should adhere to these commandments:

1. Develop a separate resume for each job opportunity, even within the same company. Keep track of which resume you sent or delivered to which potential employer.

2. Put in a job objective sentence at the top, right under your name and address. Put a single paragraph summarizing your skills and experience right under the job objective sentence.

3. Keep your resume to two pages maximum. One page is better, unless you have really stupendous qualifications for the job.

4. Make your resume easy to read; keep it clean and use wide margins and good-quality white paper.

5. Make sure strengths relevant to *this job* are highlighted, rather than merely listing your previous work experience.

6. If you have had *any* high-tech exposure, put it in, using the proper buzzwords. That is, if you've used a PC, talk about being familiar with small systems. If you've taken a course, get it in, in five to ten words.

7. If you've been working for a few years, do a chronological resume, one that lists your work history starting with the most recent job and working backwards; use snappy, positive words that the hiring manager will be likely to understand. Never use jargon from your present line of work.

8. If you're relatively new to the workforce, use the functional type of resume: what your skills are and what you've done with them, whether in a work or other setting.

9. Never use colored paper, perfumed paper, or a resume cover; never include a photo of yourself no matter how attractive you are. Never get fancy with italics or unusual fonts; it's distracting.

10. Don't mention high school, salary, family, divorces, or hobbies (unless you find out that you share one with the hiring manager).

It's often helpful to have a brief cover letter to accompany your resume. You should always include a cover letter when you send your resume through the mail, even when you're sending it at the behest of a particular person whose name is on the envelope. Otherwise you run the risk that your resume might accidentally end up on the wrong desk or on the floor of the mailroom. A cover letter also gives you an additional shot at marketing yourself. Remember to avoid going into too much detail in these letters. The cover letter's sole purpose is to serve as a lifejacket for your resume—to keep it from being submerged in the tide of paper inundating the HR offices or the hiring manager's desk. To accomplish this, the letters should be short—not more than three short paragraphs that fit on one page—crisp and to the point, and ideally have a grabber, a line that will make the reader stop and take notice.

Here are some ways you can grab the attention of the HR person in a cover letter:

- **Mention the name of someone in the company.** It might be someone you met, or it might be an officer who was mentioned in the press recently (you can get this information by checking resources at your local library). An example:

  > I am very interested in working for ABC company since reading the speech given by John Markham [the president] at the Chamber of Commerce meeting last July. ABC company seems to be at the leading edge of multimedia technology, and I would like to use my five years of PR experience in this exciting and dynamic environment.

- **Ask for an interview to complete your research on the company.** Certainly, if you sit back and wait, you might get a call for an interview but being a little aggressive can't hurt. An example:

  > I've learned a great deal about ABC company's general business through researching the Internet and the business press, and I know your new line of disk drives is enjoying great acceptance in the market. I am convinced that my writing skills can help increase sales through more powerful marketing communications and would appreciate the opportunity to meet briefly with someone in your marketing department so I could present my experience and see how it can fit in.

On letters like this you should follow up with a phone call to HR and ask to talk to the person handling responses to the ad. The operator will tell you that no phone inquiries are accepted. You should respond that you have already applied and that you are following up on your written request for an interview. Operators don't like to spend a lot of time handling inquiries like this, and you are likely to end up talking to someone from the HR department. This person, of course, will not be able to find your letter and resume (unless it has been entered into a database). You should offer to send a copy of the original package. Since you now have a name of someone in HR, you should try to deliver the copy of the letter and resume personally, and parlay this into the interview you're seeking.

- **Offer to work as an unpaid intern.** If you're currently unemployed and, after researching the company, you really feel there could be a good fit, this can sometimes be a winner. At the least your resume ought to make the first cut. It can take a company weeks or even months to fill a position. Yet the hiring manager has work that has to be done now. Why not have you do it? An example:

  > I realize that you will probably have many replies to this ad and that it may take some time to review them all. The job appears so right for me that I am willing to work for ABC company as an unpaid intern, with no strings attached, until you settle on the person you want to hire.

  This approach can be more effective after you've had the first round of interviews because your statement that this is the dream match is more credible after you've actually been in the firm and met some of your future colleagues.

There are books on the market that discuss cover letters quite thoroughly, and your local library or career guidance center undoubtedly also has one or two. Remember, when it comes to your cover letter:

- Keep it brief
- Use a grabber
- Lead from strength
- Follow up with a phone call

In the two examples of resumes and cover letters that follow, it's apparent that each candidate has an insider. Jason Frank's insider, Jenny Dobbs, handed his resume to the hiring manager, Karen Sturtevant, and also learned some useful information from Jenny. He knows that Karen has had a promotion, and (though he doesn't mention it in the letter), he knows that she will be looking for a couple of project coordinators. Kami Miyakoshi-Bradley's insider is a sales rep named Jerry O'Reilly. Jerry has spoken to the hiring manager about her, paving the way for her approach.

These resumes were shown to the CEO of a medium-size high-tech company. You might be interested in what caught his eye, and why he thought each candidate would be worth interviewing:

Jason Frank

- His service as an intern and as a waiter indicates he has dealt with a wide variety of people. In the CEO's words, "I like that he was a waiter—he had to earn money that summer, and he went out and did it."
- His education.
- The recommendation from Jenny Dobbs.

Kami Miyakoshi-Bradley

- "Kami seems to be really eager, motivated to jump into the position."
- The recommendation from a successful sales rep.
- The fact that she took the trouble to acquire some knowledge of technology.

Ms. Karen Sturtevant
Customer Marketing Manager
ProGalaxy Software, Inc.   November 2, 1999
20356 Las Cruces Avenue
Los Gatos, CA 95505

Dear Ms. Sturtevant:

Jenny Dobbs spoke to you about me a few days ago and gave you a copy of my resume. When Jenny first told me that ProGalaxy was expanding and would be looking for one or two people to coordinate field marketing, product development, customer service, and sales training, I was immediately interested, since working with diverse groups to make things happen is one of my strong points.

    I understand that you've recently acquired several additional responsibilities in connection with the move to channels distribution, and I would welcome the opportunity to have even a very brief meeting with you to discuss how I can take on responsibility for project coordination and management work. I can meet you at any time, including early in the morning for breakfast, if it's convenient (Jenny tells me that you are addicted to croissants, as am I!).

    I'll call your office next week to see when we can meet.

                                      Yours truly,

                                      Jason R. Frank

23412 Pine Crescent Drive
Botsford, CT 06331
203-555-3546
jasonf@infra.dig

Jason R. Frank
23412 Pine Crescent Drive
Botsford, CT 06331
203-555-3546
jasonf@infra.dig

Job Objective:
Entry-level project coordinator position in a growing software company, which will leverage my ability to motivate persons to achieve a common goal. I am especially interested in the areas of customer service and sales support.

Special Skills and Accomplishments:
I have successfully coordinated many teams in college and in my work with voluntary organizations, on both short- and long-term projects. Projects included event planning and execution, and fund-raising drives. I work well under deadlines, am detail-oriented, and am able to remain focused on priorities. My experience also includes designing and maintaining Web sites. I have excellent interpersonal skills and am a good public speaker.

Education:
B.A. in History, Lambert College, Oswego, NY, 1998.
Short courses in Basics of Multimedia, Project Management, and Windows 98.
Familiar with MS Word and Excel, Power Point, and Photoshop.

Work History:
Intern, The MacMillan Home, Riverdale, New York, Summer 1997. Created the Community Support Club to mobilize volunteer and financial support for this residential school for troubled youths. Enlisted representatives of many local groups, including the city council, fire and police departments, and community college, to create joint projects with the teenage residents of the home: talent show to raise money; setting up Web site; and TV special on the public access station. Received commendation from Mayor's Office.

Waiter, The Show Barn, Winslow, CT, Summer 1996.

Clubs, Interests:
President, Lambert College Debate Team, 1996–1998. Organized debates with teams from other colleges; led research groups.
News and Features Editor, Campus Radio WLAM-FM 1997–1998

John F. Ames
Director of Sales   June 3, 1999
Tanglin Systems, Inc.
230 Technology Way
Lexington, MA 02178

Dear Mr. Ames:

I learned your name from Jerry O'Reilly, your Chicago District Sales Manager. Jerry and I belong to the same fitness center here in Chicago, and I believe he talked to you about me during your recent sales training in Boston.

    I want to move into high-tech sales, and I know I will be very successful. I've followed the fortunes of Tanglin, Inc. for the past year and have studied your products on your Web site. The incredible growth of company intranets offers great opportunities for expanding sales of the entire line of Xinix routers. I agree with Jerry that an active outreach effort is what will produce the qualified leads your sales force can close, and I am absolutely certain that I can help boost your sales in a major way.

    Technology, and particularly networking hardware and software, is of great interest to me, and I've learned a lot about it in the past few years. If you'll give me a chance to prove myself, you won't regret it.

    Jerry says you'll be out here visiting customers the week of June 14. May I suggest that we meet for breakfast at your hotel on Tuesday the 15th?

    I'll follow this letter up with a phone call next week to confirm whether the meeting time and place are convenient for you.

                              Sincerely,

                              Kami Miyakoshi-Bradley

3210 Highland Avenue #4
Chicago, IL 60619
773-555-2001

Kami Miyakoshi-Bradley
3210 Highland Avenue #4
Chicago, Illinois 60619
Tel. 773-555-2001

Career Objective: An inside sales position in a high-tech hardware company.

Summary of Skills:

I am a dynamic, fast-paced, and articulate person with a good knowledge of operating systems (Windows 98, NT, UNIX), networking fundamentals, and various productivity applications. I've acquired this over the past four years through evening courses, while working full time. I have great people skills, have organized and run meetings, and turned unhappy customers into repeat business. An inside sales position would be perfect for me, as my experience includes a lot of phone work. Willing to relocate at my expense.

Work Experience:

1996–Present:
Customer Relations Manager, Rebotham Management Services, Chicago, IL. I am responsible for overall management of twelve large commercial properties in downtown Chicago. I work directly with tenants, city officials, and various service providers to resolve disputes and manage a staff of three.

1992–1996:
Sales Representative, Kendall Distributors, Oshkosh, WI. I was a highly successful representative for this large home-appliance distributor, visiting virtually every independent appliance reseller in a five-state area. I exceeded my quota each year by at least 20%. About 50% of my work was through phone contact, including cold calling. I was the sales representative on the teams that established sales policy, promotional programs and discounts, and new product training.

1989:
Summer Intern, Wisconsin Cheese Board, Madison, WI. Worked on projects to promote the national sales of Wisconsin cheeses.

1991:
Bicycled across the United States, starting from Boston and ending in Los Angeles.

Education:

B.Sc. in Management, University of Wisconsin (Stout)

Interests: Bicycling, aerobics, volleyball.

HANDLING INTERVIEWS

The interviews are the most important part of the hiring process. They are the culmination of all the networking, resume preparation, and telephoning that you've done up to now. Because of their immense importance, I'd like to give you an overview of what the interview process is like.

First, it is generally a two-stage process. The first stage takes the form of an initial screening. It might be an interview with the hiring manager, who will then pass you to two or three other persons—potentially your future colleagues. Or the first stage might be an interview with your inside contact, who then introduces you to members of the team. In either case, if you make the first cut, you get to come back in a week or so and meet even more people. It's not uncommon to be interviewed by as many as eight persons.

What are all these people looking for? Of course, they are looking for attributes needed for doing the job. But they are also looking to answer the question of "fit." You can have a sensational resume, with marvelous credentials and experience, but if you don't have the "fit," you won't get in. If just one member of the team you're seeking to join says (in one of those very interesting meetings after you've gone home to sit by the phone, and in which your future is being discussed) that somehow, she's not sure, but she just doesn't see a good fit between you and the team. Then other people may start to express doubts, and soon you're dead meat.

"Fit" is how people feel about you as part of their system, to borrow the lingo of the psychologist, and that in turn depends on how you feel about them. So if you have some level of discomfort toward the end of the interview process, especially if things have otherwise gone well for you, bring it out and get it on the table for discussion in a diplomatic and nonconfrontational way. It could have been something you heard or saw; maybe the job seemed to have changed in some essential characteristic from what you thought it would be. Or perhaps you thought you detected some hostility from one of the people who interviewed you. A good way to do this is to invite any person whom you're not certain about to comment on the "fit." They will appreciate it, and you will have the chance to clear up any misunderstandings and allay any doubts. This is a good way to solidify their commitment to you. Keep in mind that one or more of them may have had favorite candidates of their own for the position, but in virtually every instance they will at least admit that they could work well with you as a member of the team.

You may feel nervous as the day for the first round of interviews draws near. If so, I suggest that you do a drive-by the day before you're scheduled to have your interviews. Check the place out. If you're fortunate, you've been able to get inside already, by charming the receptionist or some other stratagem. The easiest way, of course, is to use your insider.

But if you haven't been able to get inside, at least go to the place so you know how to get there, and sit in the parking lot for a while. A friend of mine says she "just looked at the kinds of people leaving during lunch time" so that she would realize they were just plain folks, and she could easily work with them. It's just a means of relieving the tension a bit, like deep breathing.

Another general thing to keep in mind during the entire interview process: High-tech people like to see enthusiasm on the part of the candidate. They expect you to be pumped up about the prospect of joining their company, which is the best company in the world. Don't come across as subdued, grave, or ponderously thoughtful—it won't work. Show some spirit. And *don't* say anything negative about the company you're interviewing with. Sometimes people do this to show how up they are on what's going on in the industry: "Wow, I guess from what I read that [large, overbearing competitor] is really leaning on you guys these days. But I'm sure you're gonna come out all right." And if the stock has just lost 15 percent of its value, don't bring that little fact up with anyone you talk to. They already know about it. Even offering solace is a losing tack.

Now to the hiring interviews themselves. These interviews tend to fall into three categories, as far as their structure is concerned. Into which category yours will fit will depend very much on the skills the hiring manager and others have as interviewers. Companies have made a real effort to improve managers' interviewing skills, since hiring the wrong candidate can be a very costly mistake. Still, despite how many high-tech companies offer short courses to managers and individual contributors on how to conduct a hiring interview, some managers still try to wing it.

The Chaotic Interview

You arrive at the hiring manager's office on time, with hair combed and shoes shined, dressed like Katie Couric or Matt Lauer (as appropriate; you should not, under any circumstances, engage in cross-dressing) and as prepped for the ordeal as you can be. I've heard of cases where the hiring manager doesn't show up for the interview, and he or she didn't think to call you; or the hiring manager or shows up half an hour late with a colleague in tow, having a discussion, complete with private jokes, while you hover in the background, smiling weakly and wondering whether this is all a ploy to throw you off balance.

Never fear, it usually isn't. It's just a taste of high-tech chaos well in evidence. In fact, it's usually a sign that you can control the interview, if it ever gets started. Here are a few observations that will be helpful. First, it's great (in fact, essential) to establish rapport, but in the chaotic interview, if you start out discussing your mutual interest in fishing, you are liable to lose control and end up spending the hour comparing the relative merits of salmon eggs and nightcrawlers as bait. High-tech hiring managers who aren't certain of how to conduct an interview will usually take the easy way out; and whether it's Izaak Walton or Itzhak Perlman who offers an escape (and it's rarely the latter), you may be sure the manager will use precious time trying to take it. This is *your* interview, so keep the focus on the business at hand, even at the risk of appearing brusque. One more time: A degree of pushiness is respected in the high-tech industry.

You can ask lots of questions, but your choice of questions should leave no doubt in the hiring manager's mind that you understand the job, where it is situated with respect to other activities in the department or the division, and how these fit into the big picture. Of course, if you are asked a question to which you don't know the answer, don't try to wing it. There are few dummies in the high-tech industry, and your interlocutor can spot a fake a mile off. In fact, your winning strategy for the interview has to combine enthusiasm for the job with what your unique contribution will be, and what you hope to gain from the job in terms of self-development.

I suggest that you go into the interview with a written list of questions you have about the position. Don't leave without getting all the information you require. If you sense any lingering doubts on the part the hiring manager or others in the company when they interview you, don't fret. Usually these doubts are best addressed in the final phase of the hiring interview, after you've made all the rounds. Of course that means there has to be a roundup interview at the end of the day. Try your best to make this happen. By then the hiring manager may well have talked to some of the others who've interviewed you. If any of them had doubts, you probably sensed them at the time, and if you didn't resolve them then and there, now's the perfect time. In the best of all worlds, the hiring manager will say, "Oh, Tony's just jealous because he had a candidate for this job who's never gonna make it into *my* department."

The Scripted Interview

In this type of interview the hiring manager is usually where he or she should be—behind a desk—and this time with a prepared list of questions. These are the difficult questions that the manager remembers answering when applying for a job, so why not try them out on you? For example, "What is your worst character trait?" or "Can you tell me a time when you really screwed up on the job?" These are, of course, very easy to answer once you anticipate them, but you should not make them appear easy or you will disappoint the hiring manager. The killer question used to be "Why do you want this job?" Those of you who are of a certain age and follow national politics closely may recall that Senator Ted Kennedy was once asked that question on television. At the time, the job opening was for president of the United States. Kennedy wasn't prepared with an answer.

In the old days in business, candidates usually had not thought very much about the answer, but realized that to blurt out, "Because I need the money" or "I don't really want it that much, but I haven't yet heard from company X on my dream job" would probably not be persuasive to the hiring manager. So there was a lot of stammering. Nowadays, every candidate is prepped with replies that scatter buzzwords such as *challenge* and *development*, which I counsel you to avoid, as they make hiring managers feel ill. It's much better to say frankly that you know you can do the job well, like dealing with customers or engineers or whomever, can hit the ground running, see a terrific opportunity to learn, and know you can relieve the hiring manager of some of the pressures on him. These are welcome sentiments.

The Job-Focused Interview

This interview is likely to be more demanding on you than either the chaotic or the scripted interviews. It uses a technique sometimes called *behavioral interviewing*. A job-focused interview is likely to yield the most effective outcome, as the interviewers are well prepared and determined to hire the best candidate—which happens to be you. Prior to starting the recruitment process, the hiring manager and the rest of the interviewing team discussed content exhaustively among themselves. The interview will be a team effort. They have decided what the particular skills are that are absolutely necessary for the work to be done well. Each of the prospective interviewers has a particular area to probe, to see whether your background, temperament, and work history can really bring something to the job, and he or she is prepared with specific questions to which you must give specific answers.

A sure sign that you are engaged in a behavioral interview is when the interviewer seems to be asking questions that do *not* naturally arise from what you've provided on your resume. An example: "Can you tell me about a time when you faced an ethical dilemma in your work, and how did you solve it?" Another: "Can you tell me about a time when you received really negative criticism from a co-worker [or boss, or customer]? How did you handle it, and what would you do differently today?"

How do you participate successfully in such an interview? First, be relaxed. Think about your reply. If you find it difficult to reply, it's OK to say so. These questions are designed to get you to reveal the person behind the resume, so it's fine to be natural. Sometimes candidates have simply said, "That's a hard question. I honestly don't know *what* I'd do in such circumstances." At this point the interviewer can either be an SOB and keep pressing you, or if she has an ounce of human sympathy, she might say "Well, have you thought of . . . ," giving you an opportunity to react to a concrete suggestion.

The next most important thing you can do is to know as much as is humanly possible about the company, the products, the mission of the group you want to join, and the problems it faces. You can obtain a lot of this information from the Web and also from news stories or articles in the trade press. Of course, your insider, if you've got one, will also have provided you with lots of material. There is nothing like demonstrating a genuine interest in the company, and that comes across when you show you've taken the time to learn a lot about the company.

Here are some observations about a high-tech job interview from a woman who had had no previous high-tech work experience and spent the past six years of her life out of the job market while raising her two children:

> I know they were looking at four or five other people at least. I think I got the job because the business unit manager liked me. After the account manager had interviewed me she said, "Well, let me talk to David, the business unit manager, and he'll be in here in a few minutes." I remember it took forever for him to get in there. He was probably on the phone or something. I don't know. But he came in and he stayed for two hours. And someone said later that you knew you had the job because he talked to you for so long. I said, "I don't know why he talked to me for so long." We just really hit it off. He was very impressed that I had gone to the career center and printed out articles about their products. I mean, I hadn't known anything about these products before hand.
>
> I couldn't believe that people don't do that because in all the interviewing books, it says you should do research on the company and about the products. And he was so impressed it was amazing. He wanted copies of them, you know, "Where did you get these? I can't believe you have these." I had talked to a technical writer at [company] and she said, looking up information on the company—that was the best thing I could do at the career center, how many employees they had, what they did, what their products were, and I think that's basic. You have to, especially if it's something like high tech, if you're trying to break in.

Review your work experience in the light of everything your research has turned up on the company and its style of work, as well as the job content. It's really important to arrange an informational interview or two for the days before the formal hiring interview. If you go into the hiring interview process and you don't fully understand what the requirements of the job are and what the work environment demands, you're at a serious disadvantage with the job-focused kind of interview.

It takes time to execute a good job-focused interview, and as mentioned earlier you may find yourself called back for as many as eight or ten separate interviews. This can be exhausting, of course, but it does indicate that you are short-listed for the position. If you impress people favorably, even if you lose out, you'll have learned a lot, and of course you should make it plain to one and all that you are still interested in joining the team. Many times the number one candidate ends up not joining the company—she wants too much money, or something better has come along. In such a case, you'll get the call.

COMING TO CLOSURE

What are your chances of landing a particular job, once you've made it to the hiring manager's office? Even in companies that are growing quickly, the chances may be 20 percent that you'll land the position—another reason why you should never let up on your networking and job search until you've actually received a written offer.

When you do start closing in on a job opportunity, you should be aware of a few things if you are to keep your sanity. First, the process is likely to be long. Many high-tech companies are notoriously inefficient when it comes to adding a new person. There seems to be an unwritten rule that to make this happen in less than three months would cast a curse on the company. So be patient. Some people are lucky; they have a successful interview and they're on board the next week. If you've really hit it off well with the hiring manager, and he or she seems to indicate that you shouldn't accept any other offer for two or three days, that could indicate that the odds are even you'll get the call.

Much more typical is a series of delays that appear to the outsider to be absolutely inexplicable. These delays may come from the hiring manager's not being able to maintain the same level of excitement in filling the position as he or she seemed to exhibit while you were being interviewed. Sometimes hiring managers get pulled toward other concerns and simply put filling the position on the back burner. Of course, getting the job is at the forefront of your mind, so you would like to think that this desirable outcome is also at the forefront of the manager's mind as well. Usually it isn't, though, once you've left the office.

The high-tech industry is volatile. Since the process of getting applicants to fill a slot can be lengthy, it is also possible that while the hiring manager was busy making a decision, others in the organization were reallocating headcount away from that department. In other words, the job gets eliminated even as you were being considered for it. When this happens, the hiring manager is naturally loathe to drop everything, pick up the phone, and inform you. For one thing the manager still needs and wants the position to be filled, but now this requires a battle with HR, with competing departments, and with his or her manager. The process of rejustifying a need for the slot and denigrating everyone else's need, while not appearing to be greedy, demands both time and a high level of commitment. One of these may be lacking. Also, the hiring manager, being only human, does not want to be the conveyor of sad tidings and not only may he not call you, but he might never be there when you want to call him.

Another reason for delay in the hiring process is the HR function. The HR function exists in part to assist in getting good people into the organization, and for this purpose systems have been established to review resumes and screen applicants. As likely as not, however, the hiring manager ends up interviewing those people who have circumvented these procedures. HR knows this very well. Because hiring managers themselves prefer to hire referrals, there may be just a soupçon of jealousy and resentment borne by HR people against the hiring manager, and they may well fight this out, using you as the punching bag. On one side is the hiring manager who wants very much to get you on board, but the process usually requires that the offer letter come from HR, which gives that department the chance to pass on your candidacy. Since HR is not independently qualified to judge your capabilities for actually doing the work, its issues tend to deal with process and with corporate concerns about which the hiring manager may be aware, but may not be totally dedicated to fulfilling. For example, has the hiring manager leveled the job right? Are you perhaps a teeny bit overqualified? Underqualified? Have all the corporate objectives with respect to the hiring of women, minorities, the physically challenged been considered? Has the job been posted internally for the required period? And so on. Sometimes the hiring manager may end up in a prolonged negotiation with an HR person on whether and how to bring you on board.

There may also—will also—be internal candidates for the job you're after. Some of these folks have had their eyes on that job for a long time. They've been chatting up the hiring manager, working their way into his or her confidence, releasing those subtle pheromones to influence the course of events in their favor. And here you come at the last minute, threatening to destroy months of carefully constructed expectations and hopes. Better watch your back at the water fountain! So how do hiring managers behave when push comes to shove? The fact is they tend to go for the outsider, that is for you. It has something to do with the grass always being greener. A veteran HR executive complained about this:

> There's no perfect person for a job. There's a sort of disease where you have a position and you interview a lot of internal people and most of them have made a mistake at some time or other in the past. Someone comes in, in a brand new suit from the outside, and they look perfect. And you tend to give them the benefit of the doubt.
>
> I personally would always go with the person whose defect you know. Because you know there are defects somewhere on the guy from the outside. You just don't know what they are. And in today's legal situation doing reference checking in the United States is very difficult; unless you absolutely know and trust the person you get the reference from, you will always get a white bread reference. You will never really find out [what is wrong with the candidate].

I had a friend who had already been employed for many years at Sun Microsystems, and who applied for a management job that would have meant a promotion. After lengthy interviews, he was told the good news by the hiring manager—he had the job. HR would take care of the paperwork. My friend went merrily off on vacation with his family. When he returned two weeks later, he got a call from the HR representative. You guessed it—no job. While he was away someone gave the hiring manager the resume of an outsider. After a whirlwind round of interviews, the outsider got the job. Bad news for my buddy, very good news for the outsider. No one knows how the hiring manager feels.

All these currents and crosscurrents can be going on while you're sitting at home going over the interview process in your mind and wondering why the telephone doesn't ring. Apropos of this, here's a tip. When things have gone very well during the final interview with the hiring manager, and all your future co-workers who have interviewed you have assured you that they all think you're the greatest and that it would be fun and rewarding to work with you, and that they are looking forward to having you as a member of the team—when all this is in place—then you should not hesitate to bug the hell out of the hiring manager. This is no time to be coy. You want that job, and you deserve that job, dammit, and everyone agrees on that (except, perhaps, the HR folks). You won't jeopardize your situation by calling the hiring manager twice or three times a week to find out what the current status is. On one of those calls he or she will break down and level with you about the internal problems in trying to get you on board, at which juncture you can sympathize and ask if there's anything else you can do to help things along. Usually there isn't.

SUMMARIZING THE JOB HUNT

1. Remember the steps:
   - Appraise yourself
   - Adjust your attitude
   - Acquire background experience
   - Network and identify opportunities
   - Compile a resume and cover letter
   - Prepare for and obtain interviews
   - Remain patient while coming to closure

2. Start anywhere and take any job; you'll move quickly.

3. The process may take three to six months; don't give up.

4. Have fun while looking and keep your network alive.

5. Once you get on board, help other good nontechnical people by being part of a network.

5
COMPUTERS IN THIRTY MINUTES

What do you really need to know about computers, software, networks, communications protocols, and all that other stuff, in order to feel reasonably at home in high tech and come across as one of the team to your co-workers? Not all that much, really. Since you won't be designing chips or developing software yourself, you'll be heavily involved with what I can only call "human issues." If you'll recall, at the very beginning of this book, I said that the single most important skill required for success in high tech is good communications. That's because most of the problems you confront in high tech are involved with trying to get the right things done on time. That means dealing with people inside and outside the company, discussing, agreeing, disagreeing, persuading.

Developing high-tech knowledge is important for two reasons. First, it establishes your credibility with customers, suppliers, the press, or whomever you deal with outside your company. Second, it gives you the context and the language to work effectively with your colleagues inside the company.

This chapter will give you a little background in what computers are, how they work, and familiarize you with some of the strange and wonderful expressions used by industry insiders, which will make you more comfortable in the early days of dealing with your co-workers. These colleagues will not be scientists and engineers who know all the "bits and bytes." Although you will have contact with such people, as a nontechnical employee you will not be required to understand the inner workings of the microprocessor, the chip that is the heart of the computer. In fact it's fair to say that many people in the industry, including some in fairly technical positions, such as product managers, have no more than a hazy conception of the chip's functions. Nor will you be required to know the deep technology of a network. Where you, as a new nontechnical employee, may run into problems is in meetings or presentations, where you'll hear terminology that you won't understand or where allusions are made to trends in the industry of which you may be ignorant. Not that anyone is going to poke fun at you for not knowing these things; on the contrary, the people who work on the business side of the high-tech industry are usually very understanding and willing to help. Many of them started out without a lot of knowledge too.

When you first start working, however, the high-tech environment can be a bit overwhelming. Everyone other than you seems to be so familiar with the industry; they sling the jargon around with ease and hold conversations with each other that are simply baffling to you. How can you ever break into this charmed circle of insiders? One way is to learn the jargon. A young woman who started her career in high-tech marketing after doing similar work in the automotive industry stated:

> Basic knowledge of high-tech lingo is important because a lot of acronyms are used. You can find glossaries of acronyms, and there are resources on the Internet to help you, but there are so many terms that sometimes you simply have to ask for help, even though it can be a little intimidating.

One thing you should keep in mind is that much of what you hear and see is a display of surface knowledge. Regardless of how fearlessly your colleagues may flaunt high-tech talk, they are almost certainly not intimately familiar with the technology. They usually know what they need to know to do their jobs, and no more. Most of them picked up all their knowledge on the job, or through study or company-sponsored training—just as you will. It doesn't take forever to get a handle on the technology, though it may seem an almost impossible task when you first start working. I hope that reading this chapter will give you a good start as well as encouragement to keep on learning every day. Everything you absorb, starting from day one, will make you more and more comfortable with the technology and more valuable to the industry. I've italicized the terms and concepts you're most likely to encounter during your first weeks on the job, and I've tried to convey their meaning by discussing them in the context in which they are used. Where this wasn't possible, I've given the meaning in a footnote.

SOME GENERAL ADVICE

First, get started now. Even if you're only contemplating a possible career shift, make learning a part of your career plan and start implementing it. Start reading the technology section of your local newspaper, if it has one. Read the sections on technology in the major news magazines, as well as publications such as *Business Week* and *Fortune*. The Internet has some great sites where you can read impassioned e-mails and articles about the intrigues and holy wars of the industry. Just use one of the search engines such as Yahoo!, Excite, or AltaVista to get to them. Zero in on Microsoft, Sun, or Java. How can technology arouse such strong feelings?

If you have never worked with a computer, buy one. The price of hardware is so low that you can pay for a system by giving up one candy bar a day—or, if you're not into Toblerone chocolate, as I am, you can give up something else. Your computer doesn't even have to be new, though if you buy a used one, you'll need to exercise some caution. Make sure it has enough RAM—at least 32 megabytes—or that additional memory can be added. Also, if the monitor is more than five years old, make sure it doesn't flicker. If the image isn't 100 percent steady, don't buy the monitor. A new monitor is cheaper than new eyes.

There are some pretty good books out there for beginners. One you might take a look at is *10 Minute Guide to PC Computing*, by Shelley O'Hara. It covers the basics, from hardware to software to the Internet. If you can't find a copy, check the bookstores or online book dealers, such as Amazon.com or barnesandnoble.com; you will certainly find something suitable.

Once you get your PC, you should also buy one of the fat books that deals with your computer's operating system. This isn't because you need it to operate your computer; rather, it's so you can become familiar with what the operating system does and how it does it. Another step you ought to take, as mentioned in chapter 4, is joining a user group for your particular make of computer. These groups are everywhere—ask your local computer or software dealer. Members of user groups are wonderfully patient when it comes to answering questions.

The next thing to do is to get "wired," that is, get an account through an Internet access provider. America Online (AOL) or Netcom will do fine and are "user friendly," though there are many, many other providers. Surf the net; the easiest way to start is to get a friend to show you the ropes. Get on to some corporate Web sites, or home pages, and start learning about who's doing what in the industry. Most high-tech firms list all their open jobs, and these listings can make for very interesting reading.

When you are actually working in the high-tech industry, keep a notebook and write down questions as they come to you, at least to the extent possible. Otherwise you'll forget them until they return in some other context. During meetings you'll hear expressions that may leave you totally confused. Don't feel ashamed or bashful. Just make a note, and later find a friend and get the answer. When I first started in the industry I'd attend presentations and feel pretty lost. I started scribbling down some of the more puzzling terms or issues and would ask about them later. I even remember the first phrase I didn't understand: *beta version*. I found a friendly engineer and worked up the courage to ask her what it meant. (In software, it's the first version that you feel confident is good enough to send to a few selected customers for testing in the real world. The alpha version you'd probably only want to send to your worst enemy.) Which brings me to the next point: Please remember that there is no such thing as a stupid question. Or as the old saw has it, the only stupid question is the one you don't ask. When everyone around you is merrily discussing PCI buses, for example, and you don't know whether they're talking about computers or public transportation, ask!

Set aside some time each day, even if it is only half an hour, to read an industry publication or a technical specification sheet or some other document dealing with your company's business. Again, jot down concepts or terms that are giving you trouble. Then discuss them at lunch the next day; see what people have to say about them. If a topic really interests you, or if your company is heading in a particular technological direction, you can seek out your company's resident expert on the topic and at least have him or her point you in the right direction. You will have to be sensitive to the time demands on these people, but if you succeed in lining up a coffee break or lunch with them, you can quickly receive a real—and "real-time"—education that you cannot get in any course (*real-time*, in this context, refers to the state of technology at the very moment you are conversing with the in-house expert, as opposed to the inherently dated discussions that you would encounter in a textbook or classroom.) One friend of mine, who is a technical writer, takes a tape recorder to these sessions; later she transcribes the tapes. She said:

> What I learned was that there were people disposed to help me because they liked me or because they knew that they needed to deliver some important piece of information for their own particular set of needs. So a lot of the time there was the belief that if they helped me, then they would be successful and I would make them successful, and that worked out really well.

And a young man who mans a hot line for a software company observed:

> I've been really fortunate to have good people to go to, to answer questions. Of course, when you ask the same question two or three times, eventually you do learn it. And so you ask bigger questions. Of course, if you ask the technical people questions you could research yourself, you start taking advantage of that relationship, and then it's not so easy to get answers. Building relationships and really listening and learning and trying make a big difference—then they respect you and take more time to really teach you.

There are five areas in the high-tech industry it will be helpful for you to be conversant with:

1. Computer families
2. Networks
3. Operating systems
4. Microprocessors
5. Industry jargon

If you have some familiarity with these topics as you start off, you'll feel more at home in the industry and more comfortable in your first days on the job. But you'll feel very uncomfortable if you set yourself the goal of attaining in-depth knowledge right off. It just won't happen, and you'll start to feel discouraged. I guarantee you that within six months of starting off in your high-tech career you'll be surprised (and your friends outside the industry will be amazed) at how much new knowledge you'll have absorbed.

A BRIEF HISTORY OF COMPUTER FAMILIES

A few years ago it was fairly easy to distinguish a few discrete categories that included most types of computers. These were *mainframes, minicomputers, workstations, PC's,* and *laptop* and *notebook* computers. Of course, these didn't all come into existence at the same time. First came the mainframes, the great product of IBM in its glory days. Mainframes are huge machines that have to be kept in very large, specially air-conditioned, dust-free rooms. They're tended like demanding gods by dust-free acolytes. The early machines were surrounded by all sorts of whirring and clicking *peripherals*, such as tape and disk drives (and in the really early days, paper punch-card feeders) that fed data into the mainframe, and printers that received the end product of the computer, printing it out onto *hard copy*. Whole specialized data processing departments grew up in large corporations with specialists who knew how to use the mainframes. These departments developed specialized *applications software*, or *apps*, to solve the particular problems of the business, be it scientific research, processing of loan applications, developing actuarial tables, handling personnel records and payrolls, determining the optimal time for planting bananas, or whatever. Data processing department employees would receive requests from the company's other divisions for assistance in solving problems. They would work with employees of these other divisions to scope out the problem and reduce it to terms the computer could understand. Back in the data processing department, they would place the problem in a queue to be worked on by the computer when its scheduled time came up; this was known as *batch mode processing*. (Getting time on the mainframe was often quite an issue in those days.) The data processing people would feed the next problem to the computer, get the answer, and report on it to the "customer" within the company.

So computing, in the early days, was defined as a centralized function known as *MIS*, for "management information system," within a company that was rich enough to afford the machine, the peripherals, the *programmers* to develop the specialized software, the contracts for maintenance to keep the machine going, and the managers to keep the whole department going. All this was roughly what computing was like in the 1960s. It placed tremendous power in the hands of the managers of the data processing centers because they could determine priorities and actually shape the information that the rest of the company would have to rely on for its operations. No one could question these managers because no mere mortal could understand what was going on behind those dust-free doors.

Decentralization of Computing

A major change occurred in the early 1970s. A few bold engineers in a new company, Digital Equipment Corporation (DEC), developed a new kind of computer, which they called the *minicomputer*. Despite its name, today we would consider those early minis to be very large indeed—perhaps the size of a side-by-side refrigerator/freezer combination—but compared to the mainframe the mini was very small. The minis were very powerful, nimbler than the mainframes, and most revolutionary of all, they decentralized computing. Instead of work having to be brought to the data processing specialists, the power of the minicomputer was now brought to the people who really needed the answers, by means of *terminals* that would actually stand on or next to their desks. The terminals were simply computer monitors, or screens, with a keyboard, so that data and instructions could be typed directly by the end user. The persons using the terminals became part of a *distributed computing* network. They could use the computer without having to kowtow to the data processing manager. You could have several people, each of whom would be using a terminal hooked up to the mini, and to each of these users it would seem as if he or she were the *only* user of the mini. That's how powerful these new machines were. Occasionally there might be a slight wait if some other user were doing a *compute-intensive* application. If you got too many users and the blips got to be too long, you could add a second minicomputer to the first—just hook 'em together and, presto, you doubled your capacity and the blips went away.

The minis had lots of features that appealed to the marketplace. First, they were cheap compared to the mainframes, not only in purchase price, but even more important in a concept called *total cost of ownership*, how much it would cost a company to keep one of the things going over its lifetime, a very important concept. Second, and contributing to the lower lifetime cost, minis had a smaller *footprint;* they still needed a cool, dust-free room, but it could be a small room. Space costs money. Next, it cost a lot less to maintain a mini; maintenance costs of mainframes were horrendously high. Many minicomputer sales were made by demonstrating to the customer that in as little as two years the cost of a brand new minicomputer system with terminals and cabling would be recouped by not having to pay maintenance costs on the old mainframe. And with a minicomputer installation you could add computing power as you needed it, instead of having to invest in a monster machine at the beginning. Having computing power at the desktop made teamwork possible between users on the system, whether in the same building, in different towns, or even in different countries. Finally, since computing was now decentralized, you didn't really need a whole specialized staff to control and manage it.

Minis got to be very popular, as you might imagine, except with managers of those central data processing departments, who began to have visions of ending up on bread lines. But for huge data-crunching jobs, there continued to be a place for the mainframes. Some customers had a mainframe for some applications and minis for other applications. Of course, since both the *operating systems,* the basic software that wakes the computer up and tells it how to run through its paces, and the basic designs and standards of the hardware, or *architectures,* were quite different for the mainframes and the minis, it was not possible to hook these disparate systems together, at least not at first. The mini manufacturers spent a lot of time and research money trying to do this because it made a nice sales pitch to be able to tell a customer that the systems would all work together. Finally, this goal was achieved, at the expense of considerable complexity in systems configuration. In other words, you had to add more specialized equipment to bridge between computers from different manufacturers, and because of differences in the operating systems used by these manufacturers, it was just plain impossible to bridge between all types of computers.

The Personal Computing Paradigm

In the 1980s came the phenomenon of the personal computer, or PC. These robust little items didn't need a dust-free, air-conditioned environment; they were cooled by internal fans and could be used in an office or in a home. And they were cheap. They initially stored data on removable *floppy disks*, so-called because the disks themselves were made of flexible material. A 1.4 MB (megabyte) disk can easily hold a book the length of this one. PCs began to be made with *hard disks* built right into them. The hard disks are really hard, as opposed to the floppies, and may be stacked on top of one another like tiny pancakes, with little spaces between them. They are mounted in sealed containers to keep out dust and other potential contaminants and today can hold data in the billions of bytes. The PCs displayed input and output on monitors, just as the terminals did for the minis (in fact, you could use some PCs as terminals for minis), and were hooked up to printers. The PC was responsible for the explosion of applications software because there are a fantastic number of these machines out in the market and the vast majority of them all run the same operating system and are based on the microprocessor chip made by Intel Corporation. The ruling operating system for PCs these days is Windows 98 from Microsoft Corporation, though there are many *legacy*, or older, PCs running earlier versions of the Windows OS. Apple's Mac OS has declined in importance as that company's market share has shrunk below 5 percent; developers are understandably wary about investing in efforts to write applications to the Mac OS for such a relatively small market segment.

The Cooperative Processing Paradigm

In the middle of the 1980s the paradigm began to shift once more. A new type of computer, the *workstation*, emerged on the scene. The workstation looked in some respects like a PC. It was relatively small and relatively cheap (compared to a mini), and it stood on a desktop. But that was where the resemblance ended. This new type of computer was fantastically powerful and could therefore run applications that were only dreams a few years previously—applications such as oil field reservoir simulations, medical imaging, telephone network management, three-dimensional design, currency trading, and software development. Its monitors had superb resolution. Workstations were also capable of *multitasking*, which meant that you could pull up two or more *windows*, or displays, on the screen, at once, and run one application on one or more screens while doing something else on another. Workstations were made to be *networked*, a feature that had a powerful effect on the way people worked within an enterprise. A network within an enterprise is known as an *intranet*. Networks linked users together; users could now share and exchange files directly, and two or more persons could work on the same problem at once, sharing results. Also, audio, still photo, and video *multimedia* capabilities were built into the computer, thereby introducing new ways of transmitting information.

One of the main features of a computer system is its ability to hold data in storage by means of its internal *memory*. This kind of memory is different from the permanent memory provided by storage devices such as floppies. Workstations had plenty of *random access memory*, or *RAM*. Random access memory is the semiconductor memory in your computer that provides temporary storage for applications programs and output from your work. When you turn your computer off, the content of the RAM vanishes, which is why you must save your work into a permanent memory device, a floppy or hard disk. Having more RAM meant that workstations could hold the ever larger applications programs that were growing up like mushrooms after the rain. Workstations also had huge storage capacities on their internal hard disks. But they could also work in cooperation with other, even more powerful computers called *servers*. Servers are computers whose purpose is not necessarily to do the computation (though some may be used for this), but to hold files and applications programs in storage until the workstation calls for them to be *downloaded* so that a user could work with them. Downloading the program doesn't take much time at all, since the servers are powerful and the workstations have plenty of capacity. This kind of relationship became known as *client-server computing*. Servers also offer special services to their clients, such as printing, file sharing, and database access.

This kind of computing got started in the technical and scientific communities in the 1980s, simply because scientists and technical users were naturally involved with extremely complex problems that demanded a lot of computing power; it was ideal to have this power available "on the desktop." Workstations stayed in the labs for quite a while, but then began to move into other professions where the problems demanded vast amounts of power. A stunning example is in the medical profession, where it is possible now for a surgeon to plan an operation such as reconstructive facial surgery, carry it out through a workstation, and observe the results, before the operation is actually performed on the patient.

After all these players—mainframes, minis, PCs, and workstations—were on the computing stage, interesting things began to happen. For one thing, the lines between them began to blur. Minis had already started encroaching on the mainframe's turf in the 1970s; in the 1990s workstations and servers began to move into the minis' space, and down to the "high-end" PC. And "high-end" PCs began to look remarkably like "low-end" workstations. In an interview, Scott McNealy, the chairman, president, and CEO of Sun Microsystems, said:

> Unfortunately, PC is a conventional terminology. But I have no idea what a PC is. . . . There are several ways I segment the industry. . . . [One is] according to where the product goes. Some product goes to the home, some product goes to the briefcase—the nomadic devices if you will. And some product goes into the office. Then down the hall you have servers.

One of the interesting phenomena in the high-tech industry is the constant shortening of *product life cycles*. In workstations, new products that were twice as powerful for the same price, or the same power at half the price, emerged from the labs every three years, then every eighteen months, and now every eight to twelve months. And there has also been a powerful impetus away from proprietary systems and toward *open systems*. In a proprietary system environment, all the computing hardware has to come from the same manufacturer so as to be able to work with all the others. In an open systems environment, computers from different manufacturers can work together. (This difference is covered more fully later in this chapter.)

For a brief period it was thought that laptop and notebook computers were about as small as computers could get. For one thing, so long as a keyboard was the main input device, human fingers had to be able to hit a key without hitting two or three others at the same time. A display had to be readable by human eyes. Memory requirements, too, seemed to present an obstacle to making useful computing devices any smaller. Apple was the first company to test this notion, with its *personal digital assistant,* or *PDA,* machine called the Newton. Data was input through handwriting recognition software, and handwriting was performed on a small screen using a stylus. Though the Newton eventually was withdrawn from the market, the idea was taken up by others. For example, both 3COM Corporation and Microsoft are among the many companies that have very workable, useful palm-sized systems on the market.

An even more amazing phenomenon has gripped the industry, however. The reduction in size of computers hasn't stopped with the PDA. Functional computers—chips, with memory and an operating system—are found in devices such as watches, "smart cards," pagers, and telephones, just to name a few examples. This has been made possible by impressive breakthroughs in chip design and operating system software. Furthermore, through an amazing software development by Sun Microsystems, called Java technology, it has become possible for any device running Java to link upwards to any other device. That is, an application running on your wristwatch could send information upwards to, say, a personal computer, or even a mainframe in your bank, regardless of the operating system. You might, therefore, have your entire banking record or health history or whatever on your wrist. Since Java technology also does away with the differences between operating systems, an application written in the Java programming language will run on, say, Windows 98, but also on any other operating system. Java is also the technology that brought animation to World Wide Web home pages—dancing icons and streaming banners, for example. But the biggest, most exciting area for computers running Java is in the consumer electronics field. Delphi Automotive Systems plans to offer a Java-based system to car makers, with voice-activated e-mail and navigation systems. According to an article in *Business Week,* Motorola expects to use Java in everything from pagers and cell phones to toasters.

Another important development has been the emergence of "wireless technology." Since even small computers such as PDAs can now exchange data with a network without the need for wires (through infrared beams and other technology), all sorts of possibilities have opened up for clever new applications.

Finally, a "computer" can exist without any display, or human interface. Luxury cars now being manufactured may contain as many as 150 computers, which will control every aspect of operating and caring for the vehicle except actually steering it and operating the accelerator and brakes. And controlling those functions, too, is possible—in accident avoidance systems, for example.

NETWORKS

Why do you have to know anything about networks? Because the whole world is in the process of being networked together, and if you're not a little bit conversant in this area, you're going to appear out of it to your co-workers.

A network is, in its most basic form, simply the hooking together of nodes, each node being a computer or terminal, so that they can all communicate together. There are various network topologies, or patterns, but these need not concern us here. The important thing is that a person or group of persons can communicate easily with others who may be ten thousand miles away, or in the next room or building, working simultaneously on the same problem, sharing information, and exchanging ideas. I asked a business manager to comment on the importance of an intranet—a network which is internal to a particular enterprise (though it may involve many far-flung locations). He said:

> Now, for example, if I am working with a purchase order, it's not really just me who's involved. There are maybe fifteen people who have to be involved—the workgroup. With client servers and networking, that's become possible—to simulate the complete workflow. Purchasing is a good example. It affects accounting, invoicing, the movement of goods out of the factory, inventory, shipping. There's a new kind of software that permits this, called *groupware*. It's software that enables a group of people to perform their normal business functions in a collaborative manner.

Computers are hooked together into networks in various ways. In a *local area network,* or *LAN,* the network is likely to be an *ethernet* connection. Ethernet is a set of standards for building a network, adopted by the Institute of Electrical and Electronic Engineers, or IEEE (pronounced "Eye-triple-E"). It's the oldest and most widely used type of LAN, having been developed back in 1973. It's necessary to have standards; without them the transmission of information might face the same kind of problem a train does when it tries to go from Poland into Russia. Russian railroad track has a wider gauge than does the rest of Europe; therefore every car of the train has to be jacked up and fitted with trucks that fit the wider gauge. A Russian engine is hooked to the train, and off it eventually goes. That might be all right for rail traffic, but it's not acceptable for electronic communications. Ethernet transmits data at a rate of 10 million bits per second.

There is more than one standard for LANs. In addition to ethernet, for example, there's the *token-ring network.* It also conforms to an IEEE standard—a different one. Token ring was introduced by IBM in the mid-1980s. It transmits data at 16 million bits per second.

The battle over standards is a fascinating aspect of the computer industry, and it pervades not only hardware and networks, but software as well. The explanation for this is simple: Those who develop products that become industry standards have achieved a competitive advantage. The others have to give up what they've been working on and sign up for the standard. You'll have plenty of exposure to the standards wars as you pursue your career in the high-tech industry.

Ethernet networks can use various kinds of cabling, from specially manufactured cable to already installed telephone lines. The newest cabling doesn't rely on electricity; *fiber optic* cable, relying on another standard known as *Fiber Distributed Data Interface,* or *FDDI* (by which acronym it is exclusively known in the industry), uses pulses of light to achieve much higher transmission speeds without the problem of external electrical interference. FDDI allows high-resolution graphics and digital video to be quickly transmitted, but it's much more expensive than transmitting "over copper," that is, by wire.

So far we've been talking about local area networks, or LANs. There's another type, called a *wide area network,* or *WAN* (rhymes with *man*), that is used to span the entire globe. In a WAN the information may be transmitted in different ways. These include telephone cables and dedicated telephone lines, but by far the fastest-growing method is microwave transmissions relayed by satellites that can handle voice, data, and video. It's not at all necessary that you understand this technology, but it is something to wonder at. You can sit at your desk in Boston and send a three-dimensional technical design, complete with voice commentary and written notes, to a colleague in Singapore. When you come in to work the next morning, he's done his work and sent the project back to you. If one of you is a night owl you can work together, exchanging ideas and testing each other's work as you go. Of course, you can include your friends in Germany or France as well.

Company Networks

Here are a couple of terms you should be familiar with. An *intranet* is a network that serves the business of a single company; it enables all employees to be in touch with one another, to share files, to have access to all the company's policies and procedures, and so forth. An intranet does not depend on the Internet; it's completely internal to the company that has set it up. An *extranet* is an Internet site on the World Wide Web that is sponsored and controlled by a company; access to the site is restricted (usually by a password) to the company's suppliers, customers, or both. This can enable the integration of a "supply chain" information source. Suppliers can check the company's stocking levels of its own product and can therefore ensure that the company won't run out. Customers can have instant access to technical information on the very latest products, submit orders, and track the progress of each order from manufacturing up to delivery. An extranet can save time and frustration all around.

The Internet

The Internet is a phenomenon that has exploded onto the scene in the last few years, which is having an effect on all our lives. The Internet is the much-touted "information superhighway"; it is a global network composed of tens of thousands—perhaps hundreds of thousands—of other networks. The Internet permits the exchange of data between any two people on the earth who have access to it. This isn't the place to go into the implications of the Internet for our daily lives, except to say that it is having an immediate and profound impact on everything we do—politics, business, personal communications, education, medicine, and all other professions. Advertising on the Net is expected to reach $7.7 billion by 2002, up from $940 million in 1997. Electronic commerce—business conducted over the Internet—could reach $37.5 billion, up from $2.6 billion in 1996. The 1996 federal Debt Collection Law required that all federal payments, except tax refunds, be made electronically by January 1999. The promise of electronic commerce has drawn entrepreneurs to the Net like ants to honey. New companies are forming to provide advertising, to manage Web sites, and to facilitate and conduct e-commerce. It's worth reading about the Internet to gain an idea, however imperfect, of the directions in which this aspect of technology is propelling us.

The most important aspect of the Internet for you is that it has opened up many new career opportunities for low-tech people—writers, artists, dancers, you name it—to get happily involved in high tech. Chapter 3 goes into a number of these opportunities. Innumerable small companies have sprung up to get in on the Internet action, and more are being created all the time. The people who start these companies seem to care less about the experience level of new employees than do the managers at companies that are only slightly older, so this could be a good place to start a high-tech career—particularly if you're from the world of the arts.

Getting back to networks in general, many of their technical elements are really not difficult to understand for a nontechnical person. And there are many opportunities to learn about these matters by taking short courses, attending seminars, reading books, or just asking someone who knows to give you a short "chalk talk." Networking is without a doubt one of the most important areas of the industry and will be in years to come, offering a tremendous area for growth as all the islands of computers in all the organizations and homes around the world get hooked together. *Network* is a great word to have on your resume, and I suggest you look into it.

OPERATING SYSTEMS

As mentioned earlier in this chapter, an operating system is a piece of software that gets the computer to perform its basic functions. For example, the operating system tells the computer to load application programs when you instruct it to do so. It manages the peripheral devices attached to the computer—tape and disk drives, the monitor, printers, the keyboard, and mouse—through *device drivers*, which are parts of the operating system dedicated to such management. The operating system also acts as the intermediary between the applications program and all the data that program might require and which is stored either in internal memory or in an external storage device. When the applications program calls for a bit of data, the operating system scurries off to find it and delivers it to the program. When the applications program comes up with a solution to whatever problem it's been working on, the program hands the problem over to the operating system, which dutifully finds a safe place, such as a hard disk, to store the solution.

Another very important function of some operating systems is that they permit the multitasking mentioned earlier; you can perform operations while the computer is working on another task you've set it to solve. The operating system also manages your system's security, limiting access to your files to those who know the correct password and, perhaps, shutting down entirely if someone tries repeatedly to enter your files using wrong passwords.

The operating system is like the stage manager of a big Broadway musical: it's always there behind the scenes, and things would grind to a halt without it, yet the audience (the users of computer systems) never think about it. The stars of the show (the applications programs) get all the applause.

Operating systems were developed by computer manufacturers and third party developers, and many different ones are in operation today. If you work with a personal computer, the chances are that it uses either Windows 95, Windows 98, or Windows NT from Microsoft; OS/2, developed by IBM; or the system used by the Macintosh (the current version is Mac OS 8). There are several others, some of which are no longer manufactured (for example, CP/M). Larger computers, likewise, have their own operating systems, some of which are designated by groups of letters, for example, AIX, HP-UX, VM, VMS, and ULTRIX. You don't need to know what these initials stand for. What is important to know is that (1) applications programs have traditionally been initially developed to run on a particular operating system, and that (2) most operating systems have been *proprietary*—that is, they were developed to run on one manufacturer's computers, and those computers only. For example, assume a software developer has come up with a terrific applications program for managing hospitals—everything from the scheduling of operations to the ordering of supplies, to the tracking of payroll, to monitoring the stocks of drugs. This program could, conceivably, reduce the costs of running a hospital by some 5 percent without affecting the quality of care. The developer has used a system running VMS to develop the hospital software, so ipso facto, this application will run on VMS machines. But the only VMS machines in the world are computers made by one company, Digital Equipment Corporation (now part of Compaq). If a hospital happens to use a different brand of computer, say, a large Data General machine, it cannot use the software developed on VMS.

Similarly, if I own software that will help me do my taxes and it was developed for Windows 98, it will not run on my neighbor's Macintosh. The applications software is not *compatible* with the operating system. Manufacturers who created operating systems in the seventies and eighties tried to develop systems that had unique attributes that would help them sell computers. This incompatibility certainly helped the profitability of these companies. Once an enterprise purchased a particular brand of computer with its operating system and then developed or bought software to run on that computer, the company had to keep buying the same brand of computers so the software would continue to be compatible with all of their computers. The enterprise was "locked in" by the operating system. If another computer company were to offer the enterprise a faster, more powerful, and cheaper machine, it couldn't buy the new computer because its software wouldn't run. The people who developed software also faced problems. Software development is a very complex, tedious, and painstaking process full of opportunities for errors, or *bugs*. Bugs ultimately have to be fixed, which may require a temporary *bug patch*, or a new, improved version—a software *upgrade*. Developers want to be rewarded for their labors by selling lots of *software licenses* (permission to use the software under certain terms and conditions), but they're faced with the question of which operating system to write the software for.

The computer manufacturers all have programs to try to lure the developers to write for their own particular operating system or systems because it is the applications that sell the computers, not the other way around. You might say that all the developer has to do is to count the machines out there and then choose the dominant operating system; that's how he can sell the most licenses and make the most money. True, this is easy to do in the personal computer world, which is why any developer in his right mind writes programs that can be executed easily on a personal computer with the Windows 95 and 98 and the Macintosh operating systems. But the situation is not so clear in the case of the hospital management software. Which manufacturer's computers dominate the hospital management market? Or the oil exploration market? Or the retail chain dry goods market?

Applications software developed for one operating system can sometimes be *ported* to another operating system with relatively little fuss, requiring only a simple *recompiling* (using a piece of software, the *compiler*, to translate from one operating system to another). Sometimes this can be done in a single day. Sometimes weeks or months of work by a whole team is required, and the developer may decide that the incremental sales opportunity (and the heavy costs associated with issuing a new *release* of the software for the new *environment*) just isn't worth the effort and expense. That's when the operating system manufacturer might step in with an offer to fund, in total or in part, the porting of the software to its systems, including offering engineers to help with the port.

Obviously, life would be simplified for the developer and for the consumer if there were only one operating system, or at least if there were some way to make applications run easily on more than one operating system. In fact an operating system called UNIX was developed in 1969 that has the capability of running on a wide variety of systems. But UNIX wasn't very *user friendly*; to use it required mastering complicated commands. Initially it had no *graphical user interface*, or *GUI*. A GUI is the system of pop-up menus and icons on the computer screen that make any software easier to use. UNIX found a home in scientific laboratories and computer science faculties, whose technical staffs didn't mind its complexity of use. Some companies developed commercial versions of UNIX, but since its widespread use would have eliminated the advantage manufacturers had in locking customers into their own operating systems, there was little incentive to promote it.

Nevertheless, the pressure from end users, and notably from the federal government, to eliminate the problem of incompatibility grew, and when UNIX began to feature GUIs and became much simpler to use, its advantage as an open operating system—one that could support multiple manufacturers' computers in local and wide area networks (as opposed to the closed proprietary operating systems)—started to catch on in the commercial market. Certain large commercial users and some government departments may have many thousands, even tens of thousands, of PCs on the desks of their employees. The majority of these run some version of the Windows operating system and use a microprocessor chip from Intel Corporation. It takes an immense investment to create a new operating system; Sun had over one thousand engineers working for a year to create the version of its operating system, Solaris, that runs on both UNIX and Intel machines. The aim of this investment is not to get our home computer to run differently. The aim is to win the huge market represented by the world's largest enterprises, including government departments and universities.

The ultimate goal is compatibility, and seamless *heterogeneous networking*—the tying together of all makes of computers, no matter whose microprocessor is inside. The most promising development in this arena is Sun Microsystems' Java technology. An application written in the Java programming language will run on virtually all existing operating systems.

THE MICROPROCESSOR, OR BITS AND BYTES

Computers run on tiny amounts of electricity that get routed through the heart of the computer, the *microprocessor*. If you could see a microprocessor blown up a zillion times (its actual size might be close to the size of your thumbnail), it'd look like a very complicated model railroad, with tracks running every which way, with switches and turntables that could reroute locomotives onto different lines, with signals to hold traffic, with tunnels, drawbridges, and so forth. Instead of running trains along the tracks, though, the microprocessor sends little bursts of electricity throughout the computer. The computer only recognizes two states of being—off and on, or, if you like, a burst of electricity or no burst of electricity. We can represent this on paper by using a zero to show when there is no burst of electricity, or a one to show when there is a burst. For example: 01. First there's nothing, and then there's a little locomotive, er, burst of electricity roaring down the track. Each one of these states of being, represented by the numbers, is a *bit*. That is, the zero is a bit, and the one is a bit.

How do we get from this simple stuff to having a computer be able to, say, print the letter *a*? Leaving out a lot of useful stuff like the operating system and word processing software, and just getting to the heart of things, we could arbitrarily string together a whole bunch of bits and have them represent the letter *a*. For example: 00000001 might represent an *a*. Or for that matter 00100101 could represent an *a*. For our purposes, it doesn't matter, as long as we're consistent. So when we press the *a* on the keyboard, that pushes a little train of electric bits down on its journey through the microprocessor: off, off, on, off, off, on, off, on. What we've done is send along an 8-bit word on its merry journey. This "word" is obviously not a word in the ordinary sense, only in the computer sense. Another name for this is *byte*, which is a made-up term for a bunch of bits, 8 bits to be precise.

So a bit is the smallest amount of information a computer can handle, but it can handle all sorts of combinations of bits. Computers started out handling 8 bits at a clip, or one byte. This is why you'll hear people refer to the really old PCs as "8-bit" machines. Later, as computers got more powerful, they became 16-bit machines and then today's standard, 32-bit machines. These machines are much faster than the old ones because the locomotives carry four bytes of 8 bits each.

Modern desktop computers are 32- or even 64-bit machines. Remember, to get the number of bytes, you have to be able to divide these numbers by eight (which I refuse to do, as it reminds me too much of fourth grade). So computers are getting faster and faster. So far, we've been looking at machines that have a single microprocessor, or chip, but today there are machines on the market that have *multiprocessing* capability because they have more than one chip. Computing problems are split up and assigned to the microprocessor that has some clear track, so to speak.

Now you have all these trains running crazily along their electronic tracks carrying their information through the microprocessor. Do they just all follow the same path? Just as with the model railroad, or a real railroad for that matter, there are *instruction sets* that provide traffic signals and directions for the little locomotives. Just as well, too, given the speeds at which this stuff is going. Remember that light travels at 186,000 miles per second, which means that's the theoretical speed of our little electronic locomotives. If the journey these little bits and bytes had to make was 186,000 miles long, our electronic train would complete its journey in about one second if we ignore friction and slower switching speeds. But the chip is tiny, as we've seen. So each bit of information gets through its journey in millionths of a second.

Of course, processing speeds are never fast enough to satisfy the hardware designers. Originally the instruction sets allowed for every conceivable movement of the electronic train that might possibly be called for by any kind of program. So you got very complex instruction sets to govern what would happen. These were called *CISC* (pronounced "sisk") for "complex instruction set computing." Each little locomotive had to look at the whole instruction set each time it started out, to be sure it didn't miss something important—and that took time.

Then some genius discovered that 80 percent of the work done by computers used only 20 percent of the instruction set. The other 80 percent of the instruction set was used only 20 percent of the time, but it was still there and effectively operating as a drag on the whole system. So the idea was born of eliminating this 80 percent altogether. You'd give up some functionality, but what you'd get would be a fantastic increase in the speed of the machine, simply because the locomotive didn't waste time reading stuff it would almost never need. And so the *RISC* chip was born, which stands for "reduced instruction set computing."

Now the trains were really careening down the track. Of course that still wasn't fast enough. All computers have *clocks* inside them to tell the computer when to release the next set of instructions that'll guide the packets of electrons. For example, a PC might have a 400 megahertz

(Mhz) clock speed. Now don't panic. *Mega* means million, and it comes to us from the Greek for *large*. Heinrich Rudolph Hertz was a German physicist who first demonstrated the existence of radio waves. If your PC runs at 400 Mhz, that means that every 400 millionths of a second, a new instruction (and in powerful computers more than one instruction) is released to guide those little trains on their way. Of course, computer designers soon discovered that if you could turn up the clock speed to, say, 500 Mhz or 600 Mhz or higher, you could squeeze even more blinding speed out of the systems. So that's what they're doing.

Unfortunately, computer speeds depend on more, ultimately, than just the clock speed. Remember that we're talking about speeds inside the chip. To get anything to happen, other parts of the computer must be brought into play, and these too have their effect on speed—for example, other kinds of chips, such as *memory modules*, where information is stored during an operating session and brought out when it has to be processed. And you've got a special memory hidey hole called a *cache* , which is a temporary storage place not too far from the chip, where your computer puts information it'll probably need in a hurry but doesn't want to go all the way back to the memory module to get.

Your computer also has input devices (keyboard, pen, scanners, etc.) and output devices (screen, printer, speakers), memory storage (hard disk or floppies or both), and, of course, it has a *power supply* , which takes the 110 volts from the outlet and reduces it to the tiny trickle that computer innards can digest. All these things have to be hooked together, and for this purpose, there's a special wiring system called a *bus* . But this isn't just a simple wire. A computer bus takes the electrons and sort of slingshots them along their path whenever the little things show signs of getting tired. In other words, the bus is also very important to the total speed of the computer, and a lot of important advances have been made in this part of the technology.

I can hear you say it: "Why this terrible emphasis on speed? Who gives a darn if an operation take a few seconds longer?" A lot of problems that computers help us with are extraordinarily complex and require lots of time to process. If a computer is performing *real-time* control of the emissions of a chemical plant, for example (meaning that it senses and reacts virtually instantly and predictably to data it receives), and an explosion is about to occur in .001 seconds unless some valve is turned, and there are millions of bits of information being processed each second, then speed becomes critical. Similarly, some problems presented to a computer are so complex that they require the constant manipulation of incredible amounts of information; an example would be downloading and displaying a color movie or displaying a complex three-dimensional image on a screen, using a color scanner as an input device. That tremendous processing speed also becomes necessary if several programs are running at once, or if many people are using the same computer, connected to it by separate terminals.

Are there limits to the speed at which these machines can operate? The best answer is, theoretically, yes. But it's probably not wise to make a wager on where the limit is. Technology has a funny way of surprising people.

HIGH-TECH JARGON

Every industry has its jargon; this is what shows the world that you "belong" in the industry. Technical terms are a little different; a command of such terms indicates a command of the technology, but to be one of the gang, you have to understand the jargon. To advance your learning, you should try to master what lies behind the technical terminology. If you have had no previous exposure to high-tech language, you may be in for a shock. High-tech people look like you and me, but they don't talk like you and me. I asked an artist friend of mine to read and comment on a paragraph from an old technical manual put out by the company I work for, Sun Microsystems. Here is the paragraph:

> In order to provide an efficient multithreaded kernel, free from deadlocks and starvation, many data structures and algorithms have been redesigned. Hundreds of synchronization locks have been added to the kernel to protect and arbitrate access to critical data structures. These locks use the indivisible test-and-set instructions provided by the . . . architecture. . . . Interrupt levels are no longer used to provide mutual exclusion.

And here are his comments:

> *Kernel* to me brings the image of Squanto demonstrating to the Pilgrim fathers how to make popcorn. I don't find *dead-*

locks in my *Webster's*, but it has a faintly Caribbean ring. *Starvation* is self-explanatory. I assume that the reference to "data structures and algorithms" is some sort of computerese. But *synchronization locks*? Hundreds of them, added to that poor, inoffensive prepubescent piece of popcorn? Summons up grisly images of the worst offenses of the body-piercing subculture. I like "protect and arbitrate!" Perhaps things are not quite as bad as I thought. And *indivisible*—isn't that what the Pledge of Allegiance says our country is? *Architecture* gives us the comforting thought that above it all some Presence has worked it all out . . . and those nasty interrupt levels are no longer used to mutually exclude people.

It is unlikely that you, as a nontechnical person, will be required to do much better than my friend at deciphering technical literature, at least not initially, but you will not be able to escape the jargon and strange constructs that are used every day by your colleagues at work. There's the story of a young man who was hired as a contractor by a company to do administrative work in the engineering department. The first day on the job he overheard a co-worker observing that the chief of the department revered UNIX. The young man wore a worried expression for several days until someone explained to him that the reference was to an operating system, not to the guardians of the sultan's harem.

The very best source I've found for explaining technical terms is *The Computer Glossary*, by Alan Freedman. This is not just a glossary, but a whole course in high tech, disguised as a glossary. If you keep this on your desk at home or at work, it will greatly facilitate your study. It will help you get a command of the many acronyms that you'll hear every day on the job—a command that most of your colleagues do not have, incidentally. It describes everything about high tech that suits the level of understanding that you, as a nontechnical person, will require. It does so with copious illustrations, photos, and diagrams as well. The *Computer Glossary* does not explain some of the specialized business terminology, however.

Jargon is the talk that labels you as an insider. To that end, and also because it's rather weird and fun, I thought I'd share with you some of the expressions that have bemused me over the past few years. Here they are—listen for them as you go in the front door.

Directory of Computer Jargon

access *vb., trans.* To locate, discover. "I can't access your resume. I know it's got to be around here *somewhere.*"

architect *vb., trans.* To plan, to write. "Whoever architected this technical paper really knew his stuff."

beta *adj.* In software, the first version of a new product that is sent out to certain (usually large) customers for testing: "I heard that the beta version is relatively free of bugs."

box *n.* A computer.

bundle *vb., trans.* To include one product with another. "Windows NT came bundled with my PC."

channel *n.* A distributor, VAR (q.v.), OEM (another q.v.), or other reseller acting as an intermediary in moving your products to market. "Sales are down in the Balkans; maybe we should rethink our channels strategy."

configure *vb., trans.* Arrange hardware and software into a usable system; also, to arrange anything into a system. "We got a space problem here, so when you come in we'll have to reconfigure this broom closet."

CS *n.* Customer service; often includes services such as hot lines, repair, customer training on products, upgrades, and other after-sale activities.

cube *n.* Cubicle; your workspace or office in a company whose spatial philosophy includes the use of low, moveable partitions. "Jane's cube is ten inches wider than mine; how did she engineer that?"

datapoint *n.* Fact. "The third datapoint in your cover letter really touches on how we can use your background in psycholinguistics in our PR department."

down *adj.* Not working, not functioning. "The whole network was down for ten hours after the thunderstorm."

downsizing *n.* Layoffs in the industry; also the move of a customer from larger systems to smaller ones.

drag-and-drop *vb., trans.* Sounding like a dance craze from the forties, this actually refers to manipulating the cursor on the monitor screen so as to move files from one place to another; for example, from a folder to the printer.

end user *n.* The person who actually makes use of a product in his or her daily life.

EOL *adj.* End of life. What happens to a high-tech piece of hardware after a few months, whether customers are still buying it or not. Also used as a verb. "I hear that Product Marketing's gonna EOL the Viper machine next June."

focal *n.* Short for "focal review," which is in turn short for sitting down with your boss every few months to see how you've been performing.

foundry *n.* A factory where computer chips are made. Has as much resemblance to a real foundry as a Paul Revere silver bowl has to a trash can.

functionality *n.* Usefulness. Always preceded by the word *enhance.* "The new multithreading through the kernel to the shell produces enhanced multitasking functionality."

geo *n.* Short for *geography*, but used to mean an area or region of the world. "The discounts for the European geo are too high." "She's taken three weeks to tour all the geos and review their CS policies."

gig *n.* Short for *gigabyte*, or one billion bytes. "My new hard drive has 8 gigs of memory space."

goal *vb., trans.* To set a target (for someone). Usually used in the past participle form: "Joe was goaled to close three million last quarter."

going forward *adv. phr.* A harmless and virtually meaningless phrase used to end sentences during presentations or meetings, and intended to indicate the future, a mild resolve to make progress, or to do better next time. "To summarize, we'll be getting our products released on schedule, going forward."

goodness *n.* That which is desirable. "She postponed the decision on whether to make a hiring offer to Tom, and from the other candidates' point of view that had to be goodness."

granular *adj.* A gritty word, meaning "level of detail." "Your report is too general; it's gotta be more granular."

GUI *n.* Pronounced "gooey," this stands for "graphical user interface." The GUI is the system of little icons and menus on your computer's monitor screen that makes the software easier to use.

headcount *n.* An open job backed by actual appropriated money. "You'd be just right for this job. If only I could get a headcount."

icon *n.* The sole contribution of the Russian Orthodox Church to the world of information technology, this expression refers to any one of the little pictures on your computer screen that stand for various operations of the system, thus making the system usable by illiterates.

interface *vb., intrans.* To meet (always followed by *with*). "I guess you've already interfaced with the HR folks, right?"

interface *n.* See GUI.

iron *n.* A computer.

ISV *n.* Short for "independent software vendor." The ISV makes software that actually is useful to the end user, such as spreadsheets, word processing programs, tax preparation programs, and many thousands of others.

leverage *vb., trans.* To multiply an effort through the use of intermediaries. "We can leverage our distributors' sales forces to get better coverage of the market."

look and feel *n.* Not an invitation to licentiousness, this phrase refers to the unique nature of a GUI; this is patentable as intellectual property.

map *vb., intrans.* To correlate with, correspond to. "Your undeniable talents just don't map to our needs at present."

media *n.* The tangible medium in which software is contained; therefore a floppy disk, hard disk, tape, CD, etc. Always used as a singular noun; you will never hear the word *medium* (unless you work in a group of spiritualists).

migrate *vb., trans.* Gently nudge a customer from one version of what you're selling to another, sometimes improved, version, while garnering more revenue along the way. "We're gonna try to migrate the universities from the Viper DX99 to the Boomslang ZX007, 'cause the Viper's going EOL."

migration path *n.* The planned succession of products, the idea being that no product can remain static but, rather, is always in development. If a product goes EOL, the migration path is supposed to offer customers depending on it a way out.

mindshare *n.* Attention. "I've tried to get an additional headcount on my boss's agenda, but I'm competing for mindshare with the new executive bonus system."

MIPS *n.* Millions of instructions per second; a now out-of-date way of measuring the speed of a system.

mission critical *adj. phrase.* Critical. If it doesn't work, your customer's business will go belly up.

multitask *vb., intrans.* Doing lots of things at the same time. "I tried to set you up with our marketing manager, but between fighting the reorg, closing out Q1, and working on customer visits, he's multitasking today."

numbers *n.* In sales, the sales goal or target for a salesperson. "Did Jeannie make her numbers last quarter? I haven't seen her around recently."

OEM *n.* Short for "Original Equipment Manufacturer." The OEM buys what you make, and tucks it away inside something he makes. The end user isn't aware it's there (unless it breaks).

OTE *n.* Stands for "on target earnings." In the sales function, it's how much a rep gets paid if she makes her target numbers exactly (neither over nor under target).

outplace *vb., trans.* To fire someone. "Tim and Jane were outplaced last Friday."

package *n.* What the company offers you when you are outplaced. Usually two or three months' salary, plus two weeks' salary for every year you've been with the company. "Hey, I hear Alan took the package."

penetration *n.* Signifies a degree of success in the market with a product. The phallic connotation is intentional, the market being regarded by marketing folks (male and female) as a supine and willing receptacle for whatever new schemes they come up with. "We've had good penetration with our new Zotz CD drive."

reorg *n.* Reorganization of a department, division, or company, sometimes synonymous with chaos: "I hear a new reorg is coming down." Occurs frequently.

reprofile *vb., trans.* To update one's knowledge and skills. "The engineers had to reprofile themselves in a hurry to avoid being outplaced."

req *n.* Often confused with the Arabic word for "petition to the Grand Vizier," it is short for *requisition*. Formal application from the hiring manager to his superiors to approve creation of a job. "Your background in early religions of the Ainu is just what we need for our new marketing campaign; I'm going to put in a req tomorrow."

rich *adj.* Means "it's got lots of features." "The Kudzu Toolkit is a rich framework for developing business-critical services to be deployed over the network."

rightsizing *n.* Used to describe the situation where a potential customer has come to his senses and decided to throw out his costly old mainframe and replace it with your company's systems. Also sometimes used to describe your company's decision to throw out lots of costly old employees.

robust *adj.* For some reason high-tech marketeers use this word instead of *strong*.

rollout *n.* Introduction of a new product to the marketplace: "They chose the Pago-Pago Hilton for the rollout of the Zaftig-3 server."

RTFM *vb. phrase.* Stands for "read the f____g manual." A plea from your technical friend to you, after you've asked advice for the third time on something that's wrong with your system.

RTU *n.* Right to use; what you get when you buy a software license.

SCSI *adj.* Pronounced "scuzzy." This is not an insult. Instead, it stands for "small computer system interface." Very few people in the computer industry know this (or indeed the meaning of most acronyms they use), but that doesn't inhibit them from dropping them freely. "The new 750 Mhz FlashPlus comes with 64 megs of RAM, 6.8 gigs on the hard drive, a ZQ bus, and 3 scuzzy ports."

spiff *vb., trans.* To offer a monetary incentive to a salesperson. "The District Manager spiffed the sales guys $300 for every X30 box they sold last quarter."

support *vb., trans.* This verb means, in high-tech language, what it means in English, but it is always used backwards. That is, a high-tech person will say, "The new system supports both the Frammis Plus and Didgery-Doo Operating Systems," making it seem as if the new system is doing the two operating systems some sort of favor, such as making them more salable. In reality it is the two operating systems that make the computer system worth anything at all to end users. Without them, it is an expensive piece of junk.

system *n.* Always use this word when you want to use the word *computer*. To a high-tech type the "computer" is a chip somewhere inside a metal box.

timeframe *n.* Gratuitous sign of specious erudition (as is this definition) added to the name of a month. "I just got a call from my boss. Can we reschedule this interview around the September timeframe?"

timeout *n.* Sign language signifying that the person using it wishes to override your conversation with the hiring manager to discuss where they should go for lunch. A minor timeout is signaled by making a T with the two forefingers. A major timeout (e.g., where it's already five past twelve) is made with the forearms. (You will not see any of this, since it will be done behind your back—unless you are sitting in the hiring manager's chair, and he is sitting with his back to the door.)

twisted pair *n.* Sounding like a perverse duo act on MTV, this really refers to telephone wires in a building, which make one of the possible cabling systems for a local area network.

up and running *adj. phr.* Working, functioning. "My SE finally got the system up and running."

upgrade *vb., n.* To improve a piece of hardware or software by adding something to it, or by replacing it with an improved version (vb). The latest or improved version (n).

value add *n.* Contribution. Often seen in context of a query: "What's his value add to the team?" Or, sometimes, pseudo-bravado: "Let's make an extra effort to get some value add on the project."

VAR *n.* Short for "value added reseller," one of the channels used to get product into the hands of end users.

version *n.* In the software world, the latest product incorporating improvements, bug fixes, or whatever was needed to solve problems that arose in the earlier software. The version of software is indicated by a series of numbers. Thus the JDK (Java Development Kit) 1.2 is actually an improved version of JDK 1.1.

virtual *adj.* Characterizes that which isn't really there, but can be called into being when necessary, so we pretend that it is really there all along. Used for everything in the high-tech industry and in the lives of its denizens; impossible to misuse, since no one really understands it. "You never see Jenny with Tom; she's his virtual girlfriend." "I had a virtual pizza for lunch." "All I have in my purse is virtual money."

> **WYSIWYG** *adj.* Neither a Dickens character nor the Latvian Minister for Home Affairs, this acronym stands for "what you see is what you get"—i.e., what is on the computer screen is exactly what will print out.

One more thing. You'll find that computer folks, having already buried the adverb (the last sighting of an adverb was in the lobby of Apple Computer's headquarters in early 1991) are now seriously trying to restructure the English language. Educated men and women are saying such things as "Finance sent the e-mail to Harry, Amy, and I." Somehow the inclusion of a first name or two is taken to justify avoiding the word *me*. This assault on the objective case must cease. So when you hear it (and you will), please utter a loud shriek or take such other action as will draw attention to this horrific error. Just don't scream during a hiring interview; wait until you're safely on board.

6

THE HIGH-TECH CULTURE

Companies have cultures, just as do nations, tribes, and ethnic groups. When you start off in a new job in any company, you've entered into a culture that may be very different from any you've hitherto encountered. If the new culture suits you or you can easily adapt to it, your new job is going to seem a lot more pleasant. You'll find the work environment supportive and invigorating. If, on the other hand, the new culture does not suit you at all, it almost doesn't matter how well you can do your job. You're going to feel uneasy, uncomfortable, and eventually you'll leave. A woman who came to the industry from several years in another field said:

> I think that the culture, learning about that before you pick a job, is probably the most critical factor that is going to allow you to stay in that particular company for any length of time.

One characteristic of most high-tech companies is that the employees like to mix work with having fun. As one executive put it, "We work hard and we play hard." Some companies have "fun committees" whose purpose is to plan events where employees can let down their hair and shake off the stress of work. A young woman who works for a somewhat conservative hardware company observed:

> Whether or not you could say we play at the company, I do think people have a fair amount of fun there. The building I work in has a history of really doing it up at the holidays. People really deck out for Halloween, and they decorate the downstairs. Often there are social events, and they have food catered in; in the last six months I'd say we've had three or four of those.

> And they also have these off-sites, where you leave the company. They're sometimes work-related and sometimes not. Just sort of team bonding, or something like that. So with my team, we went to several movies. I think we saw the whole Star Wars series. We went out to meals, and laughed and joked around, and had a very nice time. We had a large off-site where the first day was informational learning about the other groups within the organization and what they were working on. And in the afternoon we played games, volleyball. I thought that was quite fun.

The fast pace of work in high tech demands some unplanned relaxation, so companies will sometimes announce that an entire amusement park has been rented for the day, with all rides and food free for the employees. Ethnic diversity may be celebrated by having festivals, with ethnic food, music, dances, and entertainment, on the company premises. Fridays are the traditional "dress down" days, when only the weird, or those who are expecting customer visits, will wear anything but jeans, sandals, and T-shirts.

Speaking of shirts, there's a shirt subculture in high tech; almost every occasion—such as the launching of a new product, a training session, or anything even remotely memorable—results in the appearance of T-shirt's or polo or denim shirts, with appropriate logos, decorations, cynical sayings, and the like. After a few years in the industry, you'll never need to buy another shirt.

April Fools' Day and Halloween are the two principal cultural events in high tech on the West Coast, with the latter dominating, by far.

MISSION STATEMENTS

Even if you are desperate to be employed, you should expend some effort on finding out what your prospective employer's corporate objectives are. The culture owes part of its content to the drive to carry out the mission. Top executives in most high-tech companies spend time thinking about what the company is trying to accomplish and how it is to do that; these thoughts are often expressed in a corporate "mission statement." A young woman who is the HR director for a thriving start-up said:

> A mission statement is important to me because candidates [for jobs] ask me what our mission statement is. I often only have a thirty-second window in which to interest a candidate. I need to be able to explain clearly and quickly what our mission is as a company, and say what our products are, what our markets are, and what our short- and long-term goals are. You should be able to get that over in a sentence.

When you interview for a job, you ought to be aware of the mission, which you can usually find on the corporate Web site.

ELEMENTS OF THE HIGH-TECH CULTURE

What does the high-tech culture consist of, in general? There are several elements:

1. Emphasis on risk-taking
2. Importance of brainpower
3. Informality and egalitarianism
4. Frenetic pace of work
5. Hands-off management style
6. Emphasis on constant learning and relearning
7. Career self-reliance
8. Informal styles of communication

Emphasis on Risk-Taking

High tech is a fast-moving, risky, volatile business. Start-up companies regularly fail. Perhaps four or five out of a hundred start-ups survive as long as five years. Large companies also fail: A much diminished Digital Equipment Corporation, once a powerhouse in computing, was purchased in 1997 by Compaq, a maker of personal computers that didn't even exist when DEC was in its heyday. Apple Computer almost vanished from the market, except for niche markets, before starting a comeback in 1998. IBM, which once defined the computing industry all by itself, had to reduce its workforce by half a few years ago, letting go of 200,000 employees worldwide, and "reinvent" itself in order to survive.

What makes a company fail? In the case of the start-ups, the idea comes first, and sometimes it is simply not possible to "productize" it, that is, to move it into the commercial market so that you can start earning money. Sometimes several start-ups are seeking to solve the same basic problem, or to provide a product or technology for which there's room for only one or two companies. A start-up may also founder when the folks that got it going, usually very technical people, prove to be well suited for product development work, but not for marketing or sales. Sometimes the market just changes and the perceived need for a product vanishes.

In the case of larger companies, failure often is a result of inability to perceive basic changes in the marketplace. Digital Equipment Corporation was coining money with its proprietary operating system and hardware. It completely missed the shift to so-called open systems because it was too busy enjoying the fruits of its extraordinary success. Apple stopped innovating and was likewise hooked on its proprietary system, the Macintosh. After disaster struck and Apple saw its market share start to slide into the abyss, it tried to broaden the Mac OS market share by allowing clone-makers access to Mac technology. Then it tried to shift back again and eliminated the clones. Nothing seemed to work, until the company launched the popular iMac machine in 1998. As mentioned earlier, IBM went through a severe crisis and fired half its workforce. Since then, however, it has engineered a true comeback and is certainly a strong market force today. Microsoft almost came to grief by ignoring—at first—the importance of the Internet. Then, when Bill Gates saw that his battleship was heading in the wrong direction, he gave the order to reverse course. Whatever you may think of Gates and Microsoft, the company executed superbly on Gates's wishes, changed direction totally within eighteen months, and is more powerful than ever, with its emphasis on connecting the Internet Explorer tightly to the company's operating systems.

So the environment you enter when you join a high-tech company is one that is highly adapted to succeeding in a risk-filled world. Oddly enough, tolerance for failure is one of the manifestations of this adaptability. You might discover this during the interview process: If your work history shows that you left an employer because it went belly up, that won't count against you in the least. High-tech people understand that failure isn't shameful, and they admire those who take risks.

Would you enjoy working in such a volatile environment? For many people, this is a psychological hurdle to be overcome. A career change is a big thing to undertake, and there's a natural inertia that tends to keep us plodding along in the same old rut. Our present work may not be totally satisfying, but it's safe—or, at least, predictable. To go from that environment into high tech, where nothing is predictable, can be quite a wrenching experience. Of course, just because a job appears to be predictable doesn't mean it's really secure. Buggy whip manufacturers, in the early years of the twentieth century, had no reason to suspect that Henry Ford was going to put them out of business in a few years.

What often shakes people out of their career inertia is a growing awareness that it's really not as secure now as they thought it was when they started out. Teachers get fired; actuaries get fired, so do salespeople, financial analysts, and, in fact, anyone who works for industry or government. The fact is that, with perhaps a few exceptions, there really is no lifetime job security anymore, anywhere.

As I was writing this part of the book I got a phone call from a man I know in Chicago. He's been a very successful marketing manager and marketing entrepreneur in the consumer products industry. Now he wants to switch to high tech. The reason for his call was that he had learned that several Silicon Valley companies, with which he was going to have informational interviews, had instituted hiring freezes. This set off some alarm bells, even though a freeze on hiring is a rather normal response to uncertain business prospects. He wanted some reassurance that high tech wouldn't again fall prey to large-scale layoffs, as it did in the early 1990s. He said, "I've got to be sure that when I make this [career] move it won't have that risk because I've been through that twice in the consumer products industry."

Now, if ever there were an industry that was stable, I would think it would be making and selling mass-market stuff—cereal, soap powder, mayonnaise—to the public. How can there be rapid shifts in the toothpaste industry, for heaven's sake? Well, I guess there are such shifts, since he seems to have been a victim of them. My friend, however, may not be cut out for work in the high-tech industry.

Another friend of mine summed up his approach to the volatile nature of the high-tech industry:

> I would say that it's important to have an open mind, and that kind of rolls into being ready and willing to change. This valley [Silicon Valley] changes every day and every week, and if I weren't ready for change and ready to embrace it, someone behind me would embrace it faster than I could, and then I'd be out on the street. So because things change, you have to recognize that and be ready to roll with the punches. If you can not adapt easily to big changes, then I would say you shouldn't be in this industry.

In chapter 7 of this book, I discuss the matter of how to survive in high tech and why "low-tech" people have better survival rates than the techies. Right now, though, I just want you to be aware that high tech carries some risks, as well as tremendous rewards. I think you can probably do very well in the industry indeed.

Importance of Brainpower

High tech loves smart people. That doesn't mean you've got to show up with a degree from Harvard or Stanford; in fact, I know some very successful people in high tech who have only high school degrees. But you've got to click. You've got to have a certain combination of intelligence, wit, and the ability to communicate "crisply," as they say in the industry.

The reason high tech values smart people is easily answered: Only smart people can work effectively on their own developing the technology, spotting the problems, devising solutions or "workarounds," and keeping up with where the company and its products or technology are going. There are very, very few "time servers" in the high-tech industry.

A nice corollary of this worship of brainpower is the scant attention the industry gives to one's background, race, gender, ethnic status, and other such things that are irrelevant to the success of the high-tech company. This isn't to say that there aren't prejudices, glass ceilings, and so forth in the high-tech industry—there are, as in any broad-based social grouping in this country. But they are much weaker than in any other industry I can think of. There is simply no time for this sort of thing in high tech.

Informality and Egalitarianism

Similarly, there's a sort of egalitarianism that permeates high-tech companies. In most companies everyone is on a first-name basis, from the president down to the receptionist. Years ago, when I was working for Digital Equipment Corporation, everyone referred to Kenneth Olsen, the founder and CEO, as "Ken." A good many of us had never met him, or even seen him in person, yet each felt that she or he had a personal link with him.

Egalitarianism also means no fancy offices, executive bathrooms and dining rooms, and the like. I cannot recall seeing any office in a high-tech company that had wood paneling, plush carpets, and so forth. As for executive bathrooms and dining rooms, I think that they would prove to be an unremitting source of ridicule for upper management. For that reason, I don't think they exist anywhere in high tech. Reserved parking for upper-level managers, I believe, would cause a revolt.

There's a nice story about offices, again involving Digital Equipment Corporation, now a part of Compaq. One of the most striking aspects of DEC's egalitarian culture is in the layout of its offices. Visitors to the company notice that, with a few exceptions, there are no closed offices at DEC. But the interesting thing is that the few closed offices that do exist are all inside offices; that is, they have no windows. The story behind this became a part of the DEC culture. Many years ago, the story goes, Ken Olsen was being shown around a new building. The facilities manager showed him where the majority of employees would sit—in interior cubicles. The tour leader proudly pointed out the managers' enclosed offices, each with a window wall looking out over the beautiful New England countryside. After the tour was over, Olsen was asked how he liked the building. "It's fine," he said, "but there's been a mistake." The bewildered manager asked him what was wrong. "You've got the managers sitting where the employees are going to sit," said Olsen. "In this company the employees have the windows." The building was reconfigured, and from that day on no manager ever occupied an office with a window.

Is that a true story? It hardly matters (though it is true that managers' offices at DEC don't have windows, for whatever reason). The point is that DEC's employees believed it was true. They knew Ken Olsen was looking out for their interests.

Informality reigns to a much greater degree in the high-tech industry than in more traditional industries. A young man told me about the consumer products industry he had just left for a job in high tech:

> Well, in [former employer] wearing a blue shirt was very important. Blue shirt, red tie. The role in that business was, as a corporate staffer, to convey the corporate image. That was more important than you as a person. More important than who you were, what you said or how you acted was what you represented. That's why the suit was so important. It represented a sort of stability, a corporate image, a solidity, and, frankly, an intolerance for nonconformity. . . . Conveying the established corporate rules and policies was more important than being creative individually.

In the high-tech industry, IBM was known as the most stuffy and formal when it came to dress code. After its near collapse, and after unloading half of its worldwide workforce, the company relented and allowed (by IBM standards) a considerable informality. In general, the old-line companies are more formal. They have fairly rigid ideas on how people should dress, the length of their hair, whether or not pierced noses and other visible parts of the anatomy are acceptable, and so forth. The younger companies are less formal. But there are lots of variants on this theme. It's a commonplace in the industry that software engineers, for example, are as informal in appearance as their minds are orderly. They usually don't meet customers—that is, they don't go out and sell—but sometimes their presence is required during a customer visit. A Silicon Valley firm was once hosting an important potential customer from the Far East. This gentleman was courted by a series of attentive young sales and marketing reps in power suits. When the time came for the discussion with the software engineers, the door opened and three of the scruffiest individuals imaginable entered the room. These were the software geniuses, to whom everyone deferred, including the visitor from Asia.

Making the transition from old-line industries to the informal high-tech culture can be a bit bewildering to the newcomer. I spoke with a young woman who came from an old-line consumer products company in the Southeast to work in a large Silicon Valley company:

> I'll tell you, my first introduction to [the new company] was just, I was just . . . so disgusted! I could not believe that grown-up people behaved this way! First of all, I'm coming from [her former employer], where people walk around in dark jackets. I'm also a Southerner, and we have sorts of codes for the way people are supposed to treat each other, and that was a big transition for me—just living in California, where things are just a lot different, much more liberal. And so I'm making that transition at the same time I'm coming in, and here are all these people wearing Birkenstocks and jeans. And I was like, don't you have any respect? I was really uptight.

Scott McNealy, CEO of Sun Microsystems, commented on his company's culture in this way:

> Sun Microsystems has always been a very diverse environment. The computer industry has always been diverse. For example, I don't care what jewelry you wear, or where. And the only dress code at Sun is that you must. It just doesn't matter. . . . I have a saying at Sun: have lunch or be lunch. The only thing that matters is output . . . you've got to do that while getting along with everyone and being part of the team.

Frenetic Pace of Work

High tech is perhaps the only industry in the United States where you can regularly see lower-level staff driving into the employees' parking lot at 5:30 in the evening, following an off-site meeting, to "get a few things off the desk." It's also the only industry I know of where folks working on a deadline take sleeping bags to work and curl up under their desks for forty winks, before getting back to business at two or three in the morning.

This kind of workaholism is quite prevalent in smaller companies and start-ups. It's less so in the larger companies, except where a project has to be completed by a certain time and there's no way on earth this can be accomplished. Then the troops rally around and end up exhausted but happy that they've accomplished the impossible.

One of the questions you have to ask yourself when contemplating a high-tech career is how much of your life you want to devote to work. Of course, this is a real issue for single parents and all people with outside commitments they cannot ignore. The way this question should be handled is forthrightly, during the job interview process. I know many people who simply cannot "pull all-nighters" or engage in similar extended-hours heroics because of other important responsibilities. Hiring managers are usually sympathetic to such situations, in my experience. You'll want to find out what the workload is likely to be in the department or group you're planning to enter. Ask your future colleagues. They'll tell you. I asked a young friend of mine who works for a small company to talk about the work schedule in high tech. He said:

> Generally, I'd say most people in high tech work somewhere around eight to six in functions such as marketing, sales, and finance. The people who make the product tend to have strange hours. Engineers might work eleven to eight, or eleven to nine, or whatever.
>
> The further you go up in an organization, the more hours you work, and the further you want to advance your career, the more hours you work. If you really want to rise quickly in high tech, you could work eighty hours and get promoted very quickly. But here people are paid pretty well, so there's this kind of disincentive to do that; people don't really feel like, "Oh, I've got to make it to the next level."

Telecommuting—having a system at home that is connected to the company intranet—is one possible solution to the workload problem; it enables you to put in an hour or two doing your e-mails or whatever without coming into the office.

I believe that most of us in high tech find the occasional high-pressure work drive to be fun—the sort of challenge that doesn't occur in more pedestrian jobs. The culture of high tech allows for time off after such efforts, by the way. If you've been burning the midnight oil for days on end, and the project is finally finished, no one is going to ask questions if you take a day or two off to go to the beach or the golf course.

Hands-Off Management Style

A few years ago every management guru was proclaiming the end of middle management and the advent of something called the "flattened corporation." It turns out, however, that lots of work (at least in high tech) tends to take the form of projects and gets done in teams. Middle managers have proven to be useful in keeping an eye on the progress of these teams, stepping in to facilitate when assistance is needed, and keeping upper management from making foolish mistakes that might undo lots of valuable work.

This said, there is probably less traditional middle management in high tech than in any other industry, in the sense of directing and controlling of subordinates. An important reason for this is that most high-tech companies are understaffed. Middle managers tend to get pushed into the crisis of the day by upper management. I'd guess that, at the director level (just under vice president), probably half of a person's time is spent on "special projects"—projects that aren't in the job description but are important to the company. I had a friend who was the European sales manager for a software company. He lived in London, but had to fly to California once each month, for three days, to attend meetings of the compensation committee. This committee was embroiled in devising a fair compensation scheme for the worldwide sales force; a kind description of such an effort would call it a thankless task. My friend faithfully attended meetings, while work piled up for him in London. He must have done very well because at the final meeting of the fiscal year he was elected head of the committee for the succeeding year. Shortly afterwards, he left his company.

The attention of managers gets diverted into all sorts of activities, sometimes by committee work as my friend was, by requests from vice presidents for rush information, reports, or other tasks. Because of this, they can't track closely what's going on in the teams and groups for which they have formal responsibility. But that's OK. In high tech, people have considerable autonomy in how they structure and perform their jobs. Managers may inform employees about problems and then leave them alone to solve them. Often employees are not even told about specific problems; they are just asked to go to work and report periodically on what is happening. A friend of mine is in an ISV support group in a medium-size software company. The focus of this group is nominally to help independent software developers use her company's technology to develop applications, but she discovered that many companies who contacted her actually needed more than that. They needed help in getting their applications recognized in the marketplace. And the smaller customers needed introductions to potential partners with financial and market clout. My friend started working on these problems and has been so effective that her job has completely changed from what it was originally intended to be. She's kept her manager well-informed, and he supports her completely.

Normally in high tech there is little monitoring of week-to-week performance. Employees get formal performance reviews every six months in some places, every year in others. Chapter 7 covers how to set the scene for a successful performance review. Other than the formal reviews, there are weekly staff meetings where managers invite employees to give direct reports, to tell what's going on with whatever projects they may have. Sometimes this results in helpful suggestions from other members of the team.

Not all managers in high tech are angels, unfortunately. The semiconductor companies were at one time run along almost militaristic lines, and there are still vestiges of this management style around. Here is what one woman had to say about her former employer:

> I don't know if all the semiconductor companies are like this, but I can bet they are. The culture is management by fear—who can yell louder. If you can yell, you're respected. It was just a bunch of b—s—, and I don't believe in doing business like that. I don't believe in intimidating people to do what you want to do. My philosophy of life doesn't fit into that.

And another woman in a semiconductor company had some difficulties trying to get across to a manager a point that concerned hiring new employees:

> He got up in my face. He was screaming, spitting, bright red in the face, telling me how dare I tell him that we didn't look at their skill sets. . . . He was screaming at me. And I don't like that. I told him he was a f— a— and I walked out. A corporate VP called me for that and was screaming at me on the phone, and I hung up on him. And you know, after that my job became very easy. People thought I was great. And I thought this was the stupidest place I've worked at in my whole life.

In any area of high tech other than the chip "foundries," it is virtually unheard of for a manager to yell at a subordinate. A few instances like that in a manager's file (where the reports end up after the employee goes to HR with a complaint), and the manager will be encouraged to find other employment (though the easing-out process may take a couple of years in the case of a VP).

Even the good ol' boy semiconductor industry is changing its management culture. A man who has worked for several years in National Semiconductor observed:

> We were known as the animals of Silicon Valley. Very aggressive. The orientation was to get it done right now and produce a lot very quickly. There wasn't an orientation toward quality. It was a driver culture where what's important is getting results—and quickly. Then you have cultures that are more people oriented where making sure people are happy and in a happy environment is valued. They are also productive, but the focus is on the people.
>
> Part of the process we're going through now is to become more people oriented because it had gone so far the other way that people weren't very happy here, and we were losing a lot of good people for that reason.

When you're interviewing with the people with whom you'll potentially be working, you'll want to ask about the way the group's manager interacts with the troops. Things to watch for are signs that the manager is a perfectionist, or a micromanager. When one of your future colleagues says something like, "You've got to redo it for him a dozen times till you get it right. But I'm learning a lot," that is a danger signal you shouldn't ignore. Unless you're a masochist, under no circumstances go to work for a micromanager or perfectionist. I have a friend who is such a perfectionist. She runs operations for a medium-size software company and is in charge of all kinds of systems (including compensation) that require various inputs from sales, marketing, HR, and other functions. People complain that she tries to run a clockwork operation in what is essentially a chaotic environment. And they're right—that's exactly what she's trying to do. The people who work for her try to provide her with the quality data she needs, in a timely manner, but most of the time it just doesn't work. Her insistence on perfection drives everybody crazy, except her.

Management style also includes trusting the people that work for you. Sun Microsystems, for example, has no formal procedure to track vacation time taken by its exempt employees. Nor does anyone care what time you come to work. It's all done on trust. It is not unusual for employees to come to work when they need to, and not according to some universal company rule. The result is that the parking lots start to fill up before eight in the morning and are still filled at six in the evening. Contrast this with the following statement from a person who interviewed with a large hardware company:

> I interviewed with [company] and wouldn't take the job because the manager said, "Well you really do have to be in at 9:00 A.M." And there are other places I wouldn't consider working because they paid attention to hours. With Data General and with Sun that was just not a concern. They expect people to be professionals and, in fact, overwork themselves.

And a young man who interviewed at Cabletron told me: "The hiring manager said I'd be expected to be in by eight and to work until five. These are the official hours."

I do know of a couple of managers—vice presidents, actually—who believe in management by intimidation. One of them came into his new position and, within a couple of months, had summarily fired five of his top directors, apparently so he could get the full attention of those remaining and also, of course, put his own loyalists into the openings thus created. Don't despair, though. Even where such a person rules the division or department, you can still survive and have an enjoyable career—as long as there's at least one layer of management between you and the ogre.

What do you do when you join a company expecting to work for the greatest boss in the world, the person who hired you, and right after you join up, the boss takes a job somewhere else? This happened to an acquaintance of mine who worked for a software company. I happened to be in her building and passed her office. She was in there singing away and stowing papers into various files. I asked her why she was so happy. Her response:

> This is my last day at [company]. It's the happiest day of my life. When I came in [a year ago] I was supposed to work for M— J—. But then he got a promotion after three weeks, and I found myself working for A— D—. I was really upset and went to see her. And she told me, "There are lots of people walking around out there who would love to have your job. You should count yourself lucky to be here." From that day I've been working to get out, and I finally did it.

It was a shame for that company to have lost a good employee simply because a middle manager didn't know how to handle an unhappy employee. Sometimes an undesirable consequence of rapid industry growth is that employees find themselves promoted into management positions, without having received management training.

The final word (for now) on bosses comes from a woman who is a technical writer in a hardware company. She said:

> I'm starting to be of the opinion that you find the right boss or you make the boss you have be the right boss. If you work for a dud, get him to change, or move on. My first boss was terrible. . . . [But] the second boss was a real knockout. I was empowered; I was encouraged to grow and to take risks. And my goodness, from the humble technical writer who arrived here and wrote a bunch of board installations, I was managing and doing all kinds of stuff.

Emphasis on Constant Learning and Relearning

Years ago in the typical corporate career path, you would come into some entry-level job, get some training, get moved around so that your experience broadened, and then you would enter the ranks of middle management. As the years passed, you would gradually begin to capitalize on all the wisdom you'd accumulated. Your role would begin to change; since you'd seen it all, you could bring your experience to bear on difficult business issues and resolve them. As a manager you could set directions, keep track of your direct reports, and serve as mentor to promising younger employees. The environment was collegial. You could, in a metaphorical sense, "settle back" and relax a bit from the everyday business turmoil.

In the high-tech world, the picture is somewhat different. Technology changes so fast that between a quarter and a third of everything you've learned about becomes obsolete every year. A lot of time has to be put in just keeping up with where things are going. It's still a very good idea to move from one place in a company to another—so-called lateral moves—to acquire breadth of experience. Whether you move or stay put, in high tech things are in constant flux all around you. An essential characteristic of the high-tech culture is constant education, usually in the form of courses ranging from half a day to three days. Almost all companies above the start-up size offer them. It's simply a must to enroll in as many relevant courses as you can take on.

When you move into middle management, there's no letup. You have to stay current with the technology, with the market, and with what the competition is doing. So you find yourself still enrolling in courses, to achieve at least a "high-level" mastery of these areas. As a senior manager told me:

> The fundamental concept is that the particular tactical skill that is needed for the next eighteen months will change in the next eighteen months.

This, plus the fact that companies are now bluntly informing employees that they own responsibility for their own careers, is putting incredible pressure on engineers to upgrade and change their skills. The same manager went on to say:

> If that person wants to stay on the leading edge of technology, he continually has to reeducate himself. In this company the vice presidents in charge of development have said they have to reeducate the workforce every four years. . . . Right now [facing a change in technology], it's as if we're teaching everybody to play the violin the other way around—from bowing with the right hand to bowing with the left hand. Or for a writer, maybe it's as if we'd changed all the rules of grammar.

But upgrading skills is not such an easy thing to do. One problem is detecting which knowledge is becoming obsolete, then figuring out what new knowledge has to be acquired to take its place. Managers are supposed to assist engineers (and others) in working this out, but they, too, find it difficult to divine where the technology gods want them to go. All this attention to change causes stress in the workplace, particularly in the engineering community. For the first time, careers may be at risk for reasons that have nothing to do with performance, but rather with the rapid change in technology itself. Paradoxically, the high-tech employees who find themselves the least affected by such pressures, and whose relative position in the industry is growing stronger, are the people whose basic skill sets are nontechnical but who understand the technology sufficiently to use their skills effectively within the industry. In other words, people like you and me.

Career Self-Reliance

Probably no aspect of the high-tech industry has gone through as much turmoil as has the employee-company relationship. In the early years, and probably right up to 1991, companies tried to create an atmosphere of nurturance for their employees. *Nurturance* is a term I use to encompass all the attitudes and policies that are—or were—intended to recognize the great, even unique, value of each employee to the company and the regard in which top management holds him or her. The statement "Our employees are our greatest asset" implies that we will nurture, care for, and develop our employees, and in exchange we know they will give their best efforts toward achieving our goals. Virtually every high-tech company started operations with nurturance as an important element of its culture, and as long as the company did well, the system worked and inspired fanatical devotion.

A senior manager at IBM, a company well known for its culture of nurturance, reflecting on the good old days, said:

> When you came into IBM, money wasn't so important. I used to say to people I was interviewing that, if they wanted to come here for money, they would leave for money. Our pay by itself was not competitive in the marketplace. The assumption was you're getting other things out of this job: working with really smart people, being part of the leadership in the industry, being able to say you worked for IBM.

Imagine the fantastic loyalty involved in being part of IBM in those good old days, where one could take pride in being paid less than the market rate!

The greatest manifestation of the nurturance culture was the unspoken commitment by management and employee to lifetime employment. This was strongest in the older high-tech companies such as DEC and IBM, and it was almost as strong as that of the large Japanese companies. Those days are gone forever. As an IBM executive observed a few years ago: "When you came into IBM it was a company you worked for until you retired. Today you work for it until you are declared surplus."

The bond of loyalty between corporate employee and corporation was severed in the early 1990s, not only in the high-tech industry, but in many other industries, in a wave of layoffs that sent hundreds of thousands of well-educated managers and workers into the street. IBM fired half of its worldwide workforce, or about 200,000 people. Digital Equipment Corporation did the same, laying off over 60,000 people. Mobil Oil celebrated a highly profitable year by firing several thousand employees.

At least in the case of high tech, it can be argued that there were valid economic reasons that dictated such "downsizings." Revenue and profitability per employee were way behind in DEC and IBM. To bolster the theoretical underpinnings of such mass white collar unemployment, the theory of the "flattened corporation" grew up. Technology and new working arrangements had supposedly made middle management obsolete, so middle managers could be let go.

Well, to make a long story short, the flattened corporation didn't last very long in high tech. It turned out, as I've indicated earlier, that middle management did indeed have a place in the industry. That revelation, unfortunately, arrived too late for a good number of middle-aged middle managers, who had been outside the industry for two or three years and, consequently, had been largely unable to keep up with technology and market trends. I heard one estimate that only 10 percent of those laid off by IBM ended up with jobs elsewhere in the high-tech industry within a year of their being fired. I don't believe anyone knows how many eventually rejoined the industry—perhaps a significant proportion, given the huge demand for warm bodies in 1997 and 1998.

Younger, healthier high-tech companies want to believe that the company nurturance of employees hasn't entirely vanished in the era of downsizing, but still there are no guarantees. The VP for human resources of such a company recently said:

> What was possible [years ago] in terms of protecting jobs and so forth—is no longer possible in the competitive environment we're in today. So jobs are at risk, and your job [as a member of HR] is to ensure that the interests of the majority of the enterprise are looked after. If that necessitates a reduction in force, that is carried out in a fair manner. The company culture is represented in the way people are treated.

What's the moral of all this for you, who are about to enter high tech for the first time? You've got to develop career self-reliance. What does that mean? As Betsey Collard of the Career Action Center in Cupertino, California, an authority on career development, puts it, career self-reliance is "the ability to actively manage your worklife in a rapidly changing environment. [It's] the attitude of being self-employed, whether you are in or outside an organization."

In a way, this is an unfortunate situation. Work is a very social activity, and people who've found a company and a work environment they like can't be blamed for wanting this to continue. And high-tech companies, at least, have become very aware of the tremendous costs entailed in high turnover. These companies are therefore trying very hard to rebuild employee trust. When you start your employment in high tech, you may hear a lot about being a valued member of the team and how XYZ Corporation is "employee-driven." Your particular employer may have retained the nurturance culture, but let me give you a very valuable piece of advice: Don't put your trust in the company. Put your trust in yourself. High tech has boom times, and it has slow times as well. You want to be sure that you survive them all, and that, should the worst happen, you can easily slide into other employment. For that, career self-reliance is a must—a suit of armor you shouldn't really put aside.

You should never forget the concept of career self-reliance. No one in the company is going to look out for you and your interests—except, perhaps, for your mentor—other than you yourself. You never rely on the goodwill, the previous track record, or the public professions of your employer. Things change too fast in high tech, and the people who make those professions today may be gone tomorrow. If you continually remain aware of the state of your company in the industry, and keep on learning, you will be fine.

You'd think that employees, when it's obvious that their company is sliding down a steep slope, would activate their life preservers and start executing contingency plans, but they do not. Most employees who grew up in the nurturance culture find it impossible to change their attitudes and take effective preventive action. They enter a state of denial, which is highly dangerous. Their survival plans all assume that help will be forthcoming from within their company. This is rarely the case. When a company starts a downsizing campaign, it is in response to its own survival imperative. The board fires the old management and brings in a new CEO and others who have no intention of continuing the nurturance culture and job security. Employees who fail to analyze their situations dispassionately and act accordingly only postpone the inevitable. I had a conversation with a man who had come to California from England to be the VP for human resources of a growing software company. This man had sold his home and most of his possessions rather than storing them because his plan was to stay five years in the new job. When his children would be ready for higher education he planned to return to England, cashing in on his stock options and selling his house, which was appreciating at about 30 percent per year.

After about a year his employer began experiencing financial problems and had to reduce the workforce. The VP of human resources worked to do this in a sensitive way, trying to soften the shock. Then the financial news got worse, and the VP again worked long hours trying to help those laid off to get good references and leads to new jobs. In the third wave of layoffs the VP found himself on the street. He told me:

> I guess I knew how bad things were. I was in most of those meetings [where the financial situation was discussed]. But I buried myself in trying to help others. It was inconceivable that it could happen to me because I just had too much to lose.

This admirable man returned to England with his family, where he remained unemployed for several months before he finally got a job. By looking at your career as if you are self-employed, you can protect yourself from a similar fate.

Informal Styles of Communication

I like to tell people that I'm the only person at Sun Microsystems who always answers his phone. Maybe I'm exaggerating a bit, but not much. The style among many people at Sun is to let the phone ring if you're

- on another line
- in the middle of something important
- having a doorway meeting with your neighbor
- staring out the window

If you want to use the phone to contact people in my company, you soon get used to leaving voicemail for them. Then you'll get a call back, but of course, for really effective communication, you'll have to answer your phone. Otherwise the voicemail exchanges will have to suffice. This is very common in the high-tech industry.

At Sun, and at other high-tech companies, there are voicemail people, but there are also e-mail people. Some of the very top technical people will rarely respond to a voicemail but will always respond to an e-mail, particularly one that is short and succinct, so you've got to know which approach to take with various parties.

Hallway or doorway meetings are another phenomenon in much of the high-tech world. When things are really humming you may go without seeing your boss or some crucial colleague for days. Maybe you've got three or four things to discuss with her. Suddenly, as you walk down the hallway, she emerges from the coffee room. You've got her cornered, but she keeps walking, explaining that she has literally two minutes before making a presentation to a large customer. You get a weak commitment on one of your three major issues and a vague promise of a meeting later that evening in her office. Then she's gone again. This is why high tech was the incubator for the aphorism "It's better to beg forgiveness then to court denial." Mostly we just go ahead and do what we think is right. The danger isn't really that your boss will say no when she ought to say yes. The danger is that deadlines may come and go without your ever getting in to see her.

In any discussion of communication styles in high tech it's important to address the question of writing. In college you learned the importance of writing clearly, of organizing your thoughts concisely on paper, of structuring your arguments so as to withstand close scrutiny. In the high-tech industry this generally does not apply to internal communications. The most important internal high-tech communications are verbal. The least important are written. The reason for this is simple: People believe they do not have the time to read and digest a memo.

Writing is important in several areas, however:

- Legal: contracts, memoranda of understanding and the like
- PR: press releases, internal "party-line" documents, marketing communications
- Web sites: all sorts of customer and employee information
- Technical: manuals, technical "white papers" (these are critically important)
- Human resources: employee evaluations, company newsletters, etc.

Writing is not significantly important in any other area. Business letters are virtually obsolete, having been replaced by the fax, phone, and e-mail, but the skills involved in writing well are critical to another means of communication—the presentation, which is an important part of the high-tech culture.

Applicants for positions in the high-tech industry are sometimes asked if they have experience in making presentations. Often your ideas will have to be communicated upwards to senior management or, laterally, to other groups in the company. For example, if you're in marketing, you might get fifteen minutes to half an hour to do a presentation to the sales force, explaining what your group is planning to do to help them make their numbers. You usually stand up in front of a meeting using overhead transparencies or projecting slides from your laptop computer onto a screen. Sometimes, if there's been insufficient time to put together a formal presentation, you'll find yourself doing a "chalk talk," writing directly onto a whiteboard. This is why nearly all offices in high tech come with a whiteboard on the walls; even impromptu drop-in meetings usually end up with someone scribbling something down. The same thing applies to ideas you want to share with your manager or colleagues;

usually you do an informal presentation. What you don't do is write a memo and circulate it.

Companies rely heavily on e-mail, as do many ordinary folks across the country. Every employee has a computer, or *system* (to use the industry term) on his or her desk, so everyone has access to e-mail, and let me tell you, there's plenty of it sent around. When you use e-mail in your company, you might be tempted to pay some attention to little things such as grammar, capitalization, and so forth. Though I'm in favor of all that, my advice is not to put too much effort into creating well-structured messages. Just getting the idea across is sufficient. Here's what one bemused new hire had to say about e-mail in his company:

> One thing that killed me was e-mail. I was used to writing memos [in his previous non–high-tech job] and having them read two or three times before I even sent them to my boss. I would read them and have my peers read them. We would edit each other's work to be sure it was perfect and looked nice, because people would evaluate you on that kind of stuff, on how it looked. That was very important.
>
> But here I'd get these e-mails: they're filled with misspellings; there's no punctuation . . . and I'm thinking, Who wrote this? Does anybody care?
>
> You know, I was just shocked. And it was just a huge transition for me. I have loosened up a lot. I'm a lot less tight than I was.

In some companies even e-mails are considered too time-consuming. In one of the fastest-growing Silicon Valley companies, a young channel manager observed:

> Here the most important means of communication is voicemail. It's amazing. There's a whole culture around voicemail. And meetings—we have tons of meetings. The voicemail types of commitments are about 70 percent commitments. A face-to-face meeting is about 90 percent. I don't know if you ever get 100 percent commitment because things change rapidly. So the commitment you made yesterday might be OK but not applicable today. Maybe a 100 percent commitment would be getting a req signed off so you could hire someone. But again, if you don't

hire someone right away, there's no guarantee they won't pull that thing right back the next day.

PARTICULAR CULTURES

When you're contemplating joining a company, what's important to you is the culture in *that* company, and even in the particular division or department you're applying to, not in the industry in general. In the interview process you'll want to get a brief but true picture of the environment you'll be working in from your prospective colleagues. To that end, I suggest you ask everyone who interviews you the following question: "What's it like, working here?" We covered this in chapter 4 in the section on interviewing. The replies you get to that little question can either confirm your belief that this company is going to be a dream to work for, or it can trigger some small alarm bells—in which case you'll probably want to ask a few more questions.

THE FINAL WORD ON CULTURE

For most people, becoming immersed for the first time in the high-tech culture is like jumping into a New England lake on a hot day—shock, followed by delight once you've gotten used to it. Occasionally, though, it's difficult for a person to adjust. At times managers from other companies have tried to impose a different culture on their particular departments. This is usually futile, as many high-tech employees won't put up with it. For the most part, problems arise when a high-tech employee takes a job in another industry. The spouse of a woman working in high tech received his MBA from Stanford University a couple of years ago. Failing to find a job in the San Francisco area, the husband returned with his wife to the Midwest. About a year later she returned to Silicon Valley to visit friends. She said:

> It's incredible! They [the new employer] are so out of it. Everyone is judged not on performance but how she looks, how she dresses, how much she gets involved in the local church. And they never just rear back and let go! High tech is crazy . . . but I sure miss it!

7

HIGH-TECH CAREERS, AND HOW TO ~~SURVIVE~~ THRIVE IN THEM

Once you've landed your job in high tech, you'll discover that a totally new existence has begun for you, at least as far as working is concerned. This chapter gives you the information you need to avoid some perils of the fast-moving high-tech industry, such as downsizing (or, layoffs, if you prefer), being stereotyped in your job, getting a terrible boss, and reorganization. Taking advantage of five important opportunities presented by your new work environment will help you flourish and get more out of your work experience. Some of these have been alluded to earlier, but here they are again, in detail: establishing credibility by growing the job, getting visibility through presentations, using mentors, making lateral transfers, and continuing your education. Thriving in high tech is like learning to swim. You learn and apply the basics, conserve your energy, find the favorable currents and head for them, and avoid rough waters.

Let's spend just a moment on the perils. First of all, downsizing does occur in the high-tech industry. Fortunately, the industry as a whole is growing rapidly, so if company A has a layoff, it is possible to get a job offer, and perhaps multiple job offers, elsewhere. Things might get tricky if a huge company were to lay off a large percentage of its workforce, as IBM and DEC both did in the early 1990s, and if the economy as a whole is in a slump, chances are it'll be tough—but not impossible—to secure another high-tech position.

Finally, if you're over fifty years old and you've been laid off, if you manage to reenter high tech, you will almost certainly have to settle for a job with much less responsibility (and salary) when you are reemployed. Yes, I know it's illegal to discriminate against people on the basis of age, race, or sex. Age, though, is the great leveler; you can be white, black, or green, male or female. Once you pass a certain ill-defined age line (somewhere between forty-five and fifty), it becomes tough, really tough. Oddly enough, age is less of a problem if you're moving into the high-tech industry from somewhere else, say, teaching. You may be perceived as having particular skills or experience that are hard to find. In that case, you may well win the prize. If you've been laid off, though, you are, in a sense, "used goods." That doesn't mean you don't have anything to contribute, but you may have to do so as a consultant, not as a full-time employee.

This is harsh, of course, and the way to protect yourself it is to know when things are getting rough in your company and bulletproof your job by "growing" it or move before the threat becomes real.

Being stereotyped in your job may be just as dangerous. It happens to everyone (except salespeople) who stays more than three years in the same position, doing the same old things. You might have a great reputation for doing your work well, but what if the need for your work should go away? The way to avoid this state of affairs is to grow your job or to change jobs after three years or so.

Getting a terrible boss happens occasionally. There are some duds in the high-tech industry, usually brought in by other duds. Sometimes, though, people become bad managers when they are promoted to management. It's not that they're evil or stupid (though it may appear that they're both); it's that no one ever taught them how to manage people. A very common manifestation of this occurs, as mentioned in the previous chapter, when an employee goes to her boss with some complaint or other, and is told, "You should be grateful you have a job here. There are lots of people walking around outside who'd love to have your job." Such a manager doesn't understand about communication or how to handle such situations. Everyone ends up angry, and the employee gets revenge one way or another (usually against the company, by losing interest and eventually leaving).

Burnout is a phenomenon that can affect high-tech employees. One person was in a particularly stressful job: he had to field knotty problems from the field sales staff; consult with engineers, product managers, finance, and legal people to get the solution; and get the answer back to the field—all under intense time pressure. "It was like being an air traffic controller," he reported. After doing this for a couple of years he began to experience periods of intense sadness. He withdrew from involvement with his job, while still going through the motions. One day he arrived at work and couldn't bring himself to get out of his car and enter the building. He ended up resigning and joining another company.

Christina Maslach, a professor of psychology at the University of California, is an expert on burnout. In an article in the *San Jose Mercury News*, she wrote, "Stress and emotional exhaustion are part of it, but the hallmark of burnout is the distancing that goes on in response to the overload." Lack of control over one's job, conflict between the employee's values and those of the company, and other factors can cause or contribute to burnout. The solution is probably to seek professional counseling and change jobs or employers.

A reorganization is when the company, or a division, changes its structure to meet one or another perceived challenges to its future or because it was purchased by another company (or itself purchased a company). A typical event would be when two divisions are merged. Suddenly there is a duplication of effort in the sales, operations, marketing, PR, and other areas. Duplication of effort means there are too many people, or at least, it means that by laying off a few hundred or thousand employees the work can still get done (somehow), and the company, being leaner, will appear more attractive to investors. The stock goes up, and the officers get rich.

Who survives a reorganization is, obviously, a question of some interest to the employees. If company X is acquired by company Y, people in X who hold positions similar to positions in Y are definitely threatened. They have no influence with anyone in company Y, and they may be out.

If, on the other hand, the reorganization is internal to a company, one can do a lot to prepare. It's of great importance to have already developed good relations with managers one or two levels higher; when the crunch comes there won't be time to take remedial action. Everything will be moving too fast.

STEPS TO SURVIVAL AND GROWTH

It's been said that surviving is about staying in place and hoping for the best, but that doesn't really work in high tech. If you stay in one place for too long, you'll find that the industry has swept by you. The first prescription is to grow your job.

High-Tech Careers, and How to ~~Survive~~ Thrive in Them

Growing Your Job

A phrase you'll often hear within the high-tech industry is "growing a job." To grow one's job means going beyond the job as it was described to you when you were hired. It also means acquiring new skills and knowledge that permit you to do your job faster and better, and to be on the cutting edge of whatever field you're in, whether it's technical writing, public relations, marketing, or whatever. And it means gaining visibility for yourself and establishing credibility with your co-workers and your managers.

The high-tech industry is very different from other industries in that pushing the boundaries of your job outwards is at worst tolerated, at best strongly encouraged by managers. In many old-line companies, any effort to increase the scope of one's job may be met with suspicion and hostility; the entrepreneurial employee may be figuratively slapped down and told to keep his or her nose to the grindstone, and do what he or she is being paid to do. A young woman who is an auditor in a hardware company observed:

> Typical of my experience with [former employer] were my trips down to Florida where I'd meet these guys who had been working in these plants for thirty years. They'd get a Rolex after so many years, and some of them had two Rolexes. And they're saying, here comes this little girl fresh out of college trying to tell me how to do my job. So I built up this persona for myself—the way I dressed, the way I behaved, and that gave me some sort of credibility.
>
> But here [in the high-tech company] that doesn't cut it. You've got to have good ideas; you've got to be articulate; you've got to be able to give presentations; and it doesn't matter what you look like. Which is liberating, but it makes life a lot harder. I mean you have to work for your credibility a lot more. And coming from Corporate doesn't mean a hell of a lot to people. They could spit and say, "Who's Corporate? Who cares?" So individual credibility is something you really have to work at.

The following remarks by a woman who's a high-tech marketing manager also demonstrate the different environment of the high-tech industry and serve as a neat summary of some of the reasons why it's important to grow one's job:

> I find that if you are a nose-to-the-grindstone person who just goes and makes change, that's all you do. People give you a dollar and you give them four quarters. If you do that well, then five years from now that's all you're going to be doing. You're still going to be making change because you're Bob, who makes good change. And your job is not secure. If you're not constantly pushing the mobility barrier, no one's going to push it for you. If you're just sitting in your cube and doing your thing . . . then you have a tendency to be pigeonholed.

Once again, growing your job is the first step you'll take toward developing a broader range of skills and abilities in the high-tech industry. Some industry pundits have gone so far as to observe that the very notion of "job" is disappearing in this country, with the high-tech industry leading the way. This is sometimes referred to as the "dejobbing" of America; what we have are people with sets of skills, who sometimes operate singly, but more often in teams, to solve problems and advance the goals of the business. Jobs with well-defined boundaries certainly appear to be becoming an endangered species. The implication for those employees who sit still with their "nose to the grindstone" is that the job may well move out from under them, so growing your job is often related to long-term survival. There's also the idea, alluded to in the beginning of this chapter, that doing the same old thing stereotypes you after a while; being stereotyped makes it more difficult to make a strategic move when you finally decide that's what you want to do. Furthermore, doing the same job for too long is boring and makes you a boring person.

But growing your job is also important for more pleasant reasons. In doing so you meet new people and acquire new knowledge, and you lay the groundwork for lateral moves into areas that you may find more interesting than your current work. Growing the job also means standing out from the crowd; getting noticed is an absolute prerequisite to getting healthy raises, promotions, or both. A woman who manages several persons in a training department told me:

> To establish credibility, you have to get involved in a lot of projects, and that's mostly for visibility. As you gain visibility, you start making presentations and interacting with people in meetings. Then your credibility starts to build. It's six months to a

year before it starts to build. It's not something that occurs in the first three months because [the high-tech company she works for] is so different from [traditional companies].

A man who graduated with a B.A. in political science, and who is now a product manager for a software company, said:

> I started off just taking a little project, getting to know the ropes around the company, doing whatever I could just to make this little project successful. Then I started to take on more and more projects, and as I did that I started to meet more people, get more visibility, and people started associating my name with certain products. You go after the opportunities, and if you see something that needs to be addressed, you do it.

That last line probably sums up as well as anything how one goes about growing one's job. Once you've become familiar with your basic tasks and start doing those well, you look around for what needs doing and start doing it. When to inform your manager about your new responsibilities depends on the type of manager he or she is. A general rule of thumb is to do this early on, either in a one-on-one meeting or in a meeting of your workgroup or department. One thing is certain—your manager will want to be assured that you are still successfully accomplishing the basic tasks that you two discussed when you first started out until your manager gets used to the new ones and sees their importance to his or her own success. Rule number one in managing a manager is to give the manager "no [unpleasant] surprises."

In general it is not advisable to ask permission of your manager to start broadening your job, simply because it projects the image that you yourself are not certain whether you really want to do it and that you're seeking encouragement. Also, seeking permission turns you from a professional into a supplicant—never a good role for any employee in any industry. A young woman who works in a Silicon Valley hardware company said:

> An individual should not depend on the company or their manager to grow their job. That's the responsibility of the employee. So for me it may be going to school, which I'm doing. I'm getting my master's degree and the company is paying for it

because it's in a related field. For me, it may also be making a list of professional books I want to complete. Or reading professional journals or attending professional organizations.

This woman is actually engaged in expanding her knowledge base—adding new skills—in preparation for expanding her job.

Another woman who is an engineer described how she realized she needed to start growing her job after she had been successfully working for several years. Her thoughts are pertinent to all people working in the high-tech industry:

> You really always have to be looking at yourself and thinking, OK, what is something I can add to my arsenal that will make me more valuable to this company? I know in the past I've ridden on the coattails of [a specific technology]. I became an expert in it and did nothing with it for a couple of years and I finally had to wake up [when there were rumors of layoffs]. I woke up and said, "Wow, I need to teach myself some new skills because this is not going to carry me forward." [You need to] really look at your skill set and figure how to add to it, and look to the future.
>
> Fortunately there is all kinds of excellent literature around, so that's one way to upgrade. At a company like this one there is also a wealth of expertise to be gained from the employees around you, so if you can network or meet some of the people that are involved [with a new direction], that's probably most important. If you can get someone whose possibly already involved in the new area to sit down with you for a bit and give you some clues on how you might educate yourself, that's a start.

Here's a little excerpt from an interview I did with a man who's now a product manager—a fairly technical position. He had no foundation in technology when he started out:

> I think there is more to read than you could possibly read. But [people should] try to read as much as they can and then find someone that they have a comfort level with, even when asking the stupidest questions. In my case I was lucky because everyone in the company was perfectly willing to answer a question

and never roll their eyes. But if you're in an environment where that's not the case, find someone, bribe them—whatever it takes—so that you'll feel comfortable asking them anything. Because you'll find, as you develop, that there's no stupid question. It's an adage but it's true. I can't tell you the number of times I've asked someone what a particular acronym means, assuming that everyone around the table knew, and no one had the faintest idea.

Growing your job is where you begin developing the career self-reliance I referred to earlier. This is the single most valuable skill you can have in the high-tech industry. Remember, according to Betsy Collard, it's "the ability to actively manage your worklife in a rapidly changing environment. The attitude of being self-employed whether you are in or outside an organization." Again and again one encounters this emphasis on self-reliance in the high-tech industry, and it correlates closely with the growing reliance of the industry on the motivated and skilled individual contributor (who usually participates as a member of a team), rather than the middle manager. A technical editor says:

> [My direct reports] have a lot of autonomy. We have a career path for writers and, not surprisingly, one thing that marks it is the ability to work independently. Junior writers need a lot of supervision. But as they gain experience, they can work much more independently. I have quite senior writers working for me now, and I certainly want to review all their plans and counsel them and help them make plans, but I almost never read the final product.

Here are some examples of growing one's job, to give you an idea of what actually happens. Julie had only a high school degree. Through a friend who worked in a large software company, she found out that there was an opening there for an administrator. Her whole subsequent career was based on constantly growing the job she found herself in until she grew her way to management (in eight years).

> Before I came to [company] I was doing a lot of accounting work. I got laid off because they were going bankrupt, and I really didn't know what I wanted to do. So my sister said,

> "Give me your resume and I'll give it to my friend at [company] and maybe there's some temporary work you can do there." So this person lined me up with a couple of interviews, and they offered me an admin job. And I took the admin job, working for a director. Shortly thereafter, he became the director of the marketing group, and I started to do various projects outside of the admin role because I wanted more work.
>
> My boss thought it was a good idea because there was definitely a need. Every department has a lack of resources, and there are so many projects that need to be done. I started to help out at the trade shows, at which point I was starting to do more and more marketing communications projects. So my boss switched me into being the marketing communications coordinator. And he had to go back to get another admin.
>
> My big focus in this division became marketing communications, but when our brand started to get very popular . . . I switched roles, a normal thing, and I focused on the brand and became the brand development manager. It was an opportunity for me to get more experience in brand management. . . . I'm actually doing both jobs right now.

Here's another example of growing one's job. Tony worked in the customer marketing department of a software company. His job was to assist the sales force in Asia to license software to large OEMs by negotiating complex licensing agreements; for this purpose, he traveled to Asia five or six times a year. He would explain the provisions of the agreement to the customers, calling back to the United States for legal advice (he was not a lawyer) where necessary. Tony wanted to get into sales eventually; the problem was that he was doing a great job where he was, and the Asian sales force particularly wanted him to stay in his present job. He told me:

> I was frustrated at first. Then I saw that we were getting these requests from Eastern Europe and Russia; all kinds of people wanted to visit us to learn about our products. No one at headquarters was interested because they thought there was no money in those places. We had one sales guy covering all the ex-Communist countries on a part-time basis. So I decided to invite some of these groups to visit us, and I'd set up the agendas for the meetings and host them.

Tony started to get known as the person to whom all such requests should come. When the part-time salesperson started negotiating some serious deals in that part of the world, it turned out that Tony knew many of the customers, having met them on their visits to company headquarters. He was able to get a position in the sales organization, based in the United States, but with responsibility for developing business in Eastern Europe and Russia.

Renée worked for a very large computer company as a project coordinator in the company's equivalent of a public library—the technical documentation group. She said:

> I was enjoying my job, but they were reorganizing, going to a centralized structure. I wanted to be a project leader (the next step up), but with this reorganization, I knew that wasn't going to happen. And I also wanted to get out of technical documentation and start dealing with the actual products. So I found a project coordinator position in networking communications. I would have liked to have a project leader position, but I didn't think I could make that jump because I didn't have any experience that [management] would consider relevant. So I went into this interview, and I talked to a gentleman and told him what I really want to do is become a project leader—but I want to start as a coordinator because I want to learn this business.
>
> And from day one that expectation was set. I ended up getting hired for the position. In six months . . . they promoted me to project leader.

And the marketing manager for a small company that makes and sells specialized instruments to the computer industry describes how he grew his job from product manager for a single product to marketing manager:

> I grew the job. I had been in sales, but then we had a bad couple of years and my income was cut in half. So I suggested that I move into marketing, and I became a product marketing manager. I started out with just the main line product. Then we came out with a whole new product line with a different market, different customer set. We went to the venture capitalists, and they put in lots of money. We started going to all the trade shows, spending a lot of money in promotion. I got

involved in writing the sales training materials, and then in the promotion of the product. So then I had two lines, and I was doing it all. I ended up as the company's marketing manager.

There's one danger you might face as you seek to grow your job, and that is that you might end up doing a lot of work that benefits others—your colleagues, or your boss, for example—but that does not add much to your own personal inventory of useful skills. In the high-tech industry, almost everyone has too much to do, and people will seek ways to offload some of this onto whoever appears willing and eager for more work. If this contributes to your strategic direction, adds to your contacts and knowledge (and if you don't get swamped by it), fine—take it on. If not, refuse to do it.

Making Effective Presentations

When you grow your job, you start to get involved in those areas of your division or department that involve new problems and opportunities. Because they're new, there is no well-defined approach to them, so you're really breaking new ground. Chances are you'll be asked to make periodic reports on how you are doing. If you're not asked, be sure to arrange to do so anyway. To do this you call a meeting, assemble all the persons whose work is or might be affected by what you're working on, and make a presentation to them. Use your judgment, but if it seems reasonable, have your manager attend as well, so that he or she can bask in the reflected glory of your accomplishments. (Always give your manager a private briefing beforehand so there are no surprises.)

The effective presentation is the single best tool for achieving recognition and positioning yourself for advancement in the high-tech industry. It shows that you can tackle new areas of responsibility on your own and that you can leverage resources as needed from elsewhere in the company to achieve a solution. The presentation is a subtle way of doing this because you're not directly seeking any reward; you are reporting on your project. You may invite open discussion on which of two or three alternative plans is best for solving a particular problem; you may request money or people to help you finish the job. Whatever the primary purpose of the presentation, there is always a secondary purpose: to present *you*.

Many presentations in the high-tech industry are uninteresting, too long, choked with data, and are badly delivered. So many are like this that a certain resigned tolerance has grown up. People filing into a conference room *expect* a presentation to be mediocre, and often they are not disappointed. This is too bad because whether or not a bad presentation achieves its primary purpose, it usually fails in its secondary one—that of showing that you are a smart, confident, poised character who knows the subject matter cold and who is probably destined for higher things some day. An interesting article in *Fortune* stated that "true meritocracies don't exist, and self-righteousness won't pay the bills. So you'd better learn how to start bragging." The presentation offers a nice way of doing just that.

Most presentations in the high-tech industry are put together at the last minute; it's usual to see a nervous admin running to the printer, hastily revising overhead transparencies and running off twenty, thirty, or fifty copies of the whole thing to be passed out to all those attending the meeting. (In fact, it is ironic that vast quantities of paper are used up in a high-tech presentation, given that technology is supposed to liberate us from paper flow.) One of the most engaging sights to the detached observer of important presentations in the industry is that of the presenter standing to the side of a screen, using a pointer to take his audience through a crowded slide, while the members of the audience have their heads down and are apparently attentively reading the same slide in the handout. As the slide on the screen changes, there is a sound of twenty or so individual pages being turned in as many handouts. Another hallmark of the mediocre presentation is hearing the presenter reading, word for word and line for line, the contents of each slide, expanding on a given topic as he or she thinks the situation may demand.

Giving a good presentation is not hard once you are properly prepared. Many companies offer courses in presentation skills, and it is very important to sign up for one if you have the chance. If you don't have such an opportunity in your company, perhaps you can find one at a continuing education center somewhere nearby. If you can't, here are seven suggestions for making outstanding presentations:

1. **Organize your subject.** You have to think before you start making up your slides. You want the presentation to have a beginning, a middle, and an end. When you go into an important presentation, take a one-page outline with you—it'll be a lifesaver if you lose your way. Organize it along the following lines:

    A. Introduction (What I'm Going to Tell You):

    —First point

    —Second point

    —Third point

    B. What I'm Telling You:

    —First point (expanded)

      subsidiary point one

      subsidiary point two

      (perhaps) subsidiary point three

    —Second point (expanded)

      subsidiary point one, etc.

    —Third point (expanded)

      subsidiary point one, etc.

    C. The Result of What I've Told You:

    —Where we stand today

    —The two (or three) alternatives

    —Resources needed to continue progress

    —Remaining milestones to success

    There's an old rule in consulting: Always give 'em three points. Why not two, or four? Three engages the interest more; two points invite argument, and four cause confusion.

2. **Less is more.** Everyone in the high-tech industry seems to honor this rule in the breach rather than the observance. Most high-tech presentations are too long, have too much data, too many slides and handouts, and too much on each slide. The slides don't seem ever to focus in on the point of the whole exercise. Their value is mostly soporific.

    Keep slides clean. Use bullets and keep your text to five or six lines per slide and a few words per bullet. It's great discipline to try to do this; it gets you really understanding your subject matter, and it keeps you from reading the whole presentation from the screen.

If you have half an hour for a presentation, do it in twenty minutes and then invite discussion. Your audience will be surprised and delighted and will get really involved with the topic.

3. **Maintain control.** If your audience's attention wanders, or if two people at the back of the room start a private conversation, there's a very simple technique for grabbing back control. Just stop talking and stand motionless. After about five seconds, you'll have everyone's full attention.

   If you've planned ten to twenty minutes for discussion, say so at the beginning of the presentation and then stick to it. Obviously, if your boss's boss is listening and wants to ask a question, you'll have to answer it, but avoid getting drawn into side discussions until the end.

4. **Pace yourself.** Many high-tech presentations are planned for half an hour, but the presenter arrives with fifty overheads. The audience is then subjected to an attempt to edit the presentation, complete with mumblings such as, "Naw, I don't think that'll be useful. . . ." If you know you've got thirty minutes, do your presentation at home the night before in front of a mirror, or your spouse, or the cat, but get it to thirty minutes. Twenty is even better.

5. **Be calm, and look at your audience.** One presenter at a high-tech conference had the peculiar habit of waltzing every time he began speaking. He would slide his left foot forward, then lean to the right and move his right foot to the right. Next he would bring his left foot over to the right and then slide the right foot back. There were about two hundred people in the room; those toward the rear could only see the effect this movement had on his head and shoulders, while those in the front could admire the full effect. No one paid much attention to what he was saying.

   The only way to cure a case like this is through practicing relaxation while talking and using a videocamera or friends for feedback. It's the hardest thing in the world to simply stand still on a stage or on a podium, with your arms at your sides, and talk to people.

   It's important to maintain eye contact with individual members of the audience as you make your presentation. Change individuals from time to time, of course. You can be sure that anyone you are watching will be paying strict attention to what you are saying, but the real purpose of this gimmick is to keep a bond between you and the entire audience.

6. **When things go wrong, don't panic.** There's a saying in the world of acting: The audience never knows unless you tell them. When you screw up or lose your way, it's not a big deal for your listeners unless you clue them that it's a big deal. When President Clinton introduced his health care plan to the Congress and to the nation in September 1993, no one realized that he was ad-libbing for the first five minutes. Some hapless aide had put the wrong speech on the teleprompter. Almost no one in the TV audience noticed that anything was wrong.

   When things go wrong in a presentation—and normally that means you've lost your train of thought—just stop, take a breath, take a "drink of your notes," and start in again.

7. **Announce the end of the presentation.** You can do this by saying phrases such as "finally, . . ." or "to summarize . . . ." And always thank the audience for coming.

Using Mentors

A mentor is a person who helps you to be at ease with your environment, your colleagues, your work, by listening sympathetically and by giving you advice from time to time. Almost every person interviewed for this book mentioned the importance of having a mentor. Some high-tech companies even have formal mentoring programs, something that does not exist in other branches of industry. Most people mentioned having mentors inside their company, but some turned to people in companies they had formerly worked for. They seek the perspective of people who have wrestled with similar types of problems with whom they have established a deep bond of trust.

Why mentoring is so important in the high-tech world that it is so often institutionalized is a subject for conjecture. I believe that, like so much else that differentiates the industry from others, it is related to the fast pace and erratic path that technology takes, and to the fact that entire large organizations are trying to react in real time to the "paradigm shifts." Mentors provide support, encouragement, and understanding, without having anything to gain other than the good feeling that comes from helping you over the rough spots.

A woman who has been in the same company since she started working six years ago had this to say about mentors:

> There are sort of the official mentors, and there are unofficial mentors, or people who take you under their wing. The official mentors do things like show you where to get forms and all that sort of thing. But that's not necessarily what's important. The individuals who have sort of figured how to work in [company], how to get things done, are the most helpful.

Making Lateral Transfers

A lateral transfer is a move to another job within the company without any change in level, and with no promotion involved. The employees in the high-tech industry are like those of other industries: they expect that hard work and outstanding performance will result in a move up the corporate ladder. Lateral moves within the company (or to another company) were traditionally perceived as fruitless, a waste of energy, even an admission of failure. That perception has changed. Lateral moves are a great way to build a broad and deep range of experience. This gives you more latitude in case you must find another job quickly. Even in boom times, companies sometimes unload large numbers of employees. Witness Netscape Communications's struggle to survive. In January 1997, the company, despite being an industry innovator, laid off three hundred workers. Intel Corporation planned to lay off three thousand people in the spring of 1998. Sun Microsystems reorganized its software sales structure in mid-1998, and many sales, marketing, operations, and support people hit the street.

As I've said earlier, employees with a broad range of skills and experience can find new work much more quickly than those who have been pigeonholed in the same job for many years. Also, moving from your job to elsewhere in the high-tech company recharges your batteries with brand new challenges.

Many high-tech industry people today feel frustrated that they have not been promoted to "the next rung." If that's really important to you, you may have no alternative than to change companies. That's because the number of people admitted to higher levels in the corporate structure is dictated from way up high in the company. Only so many job slots are allowed at each level. Thus, although everyone may agree that you are the top performer in your group, and everyone else in the group is, say, a level eight, you may still be stuck at a level six position. Perhaps, you really should be concerned with how much money you are making and how much your contribution is valued by your manager and your team. It's easier for a manager to award an outstanding employee more salary and a bonus than it is to "relevel" the job.

Also keep in mind that, while it's true that higher-level managers, starting at the level called "director" in the high-tech industry, have bonus plans that individual contributors do not, it's also true that directors and higher-level managers do not have greater job security; they can and do get downgraded and even fired. They must attend regular meetings that can seem interminable; they may have to sit through the whole day while you as an individual contributor can come in, make a twenty-minute presentation on some business problem, and leave everybody impressed and refreshed by your command of the issue. It is not pleasant to be one of a few hundred directors and vice presidents in a large corporation and learn from a memo out of the president's office that, come next January, there will be 50 percent fewer directors and VPs than there are today.

Lateral transfers are usually very rewarding in terms of renewed work satisfaction. The following comments came from a woman who was reviewing her career at her former employer, a hardware company; she had just accepted a job in a smaller software company:

> I happen to need challenges in large doses on a regular basis, and I need variety and change. So I learn things and I get bored very quickly.

She had had four jobs in about seven years and was now seeking an environment that would offer her even more varied opportunities.

I do not mean to imply that one comes into the high-tech company and has to remain at the same job level forever. Changing jobs within a company does occasionally offer the opportunity for a promotion, and even within your existing workgroup, if you've grown your job and let your manager know about it, your job can be releveled, thus resulting in a promotion for you. When this happens it's great for you because there is no competition for the higher-level job—it's yours already. When someone in a job leaves the company, however, and you want to go after the vacant position, you are likely to discover intense competition from several other persons. Even if it's the same level for you, it may represent a promotion for them. You'll have to sell yourself to the hiring manager just as you would if you were coming from the outside, except that you now are an insider with a track record, and you have people who will go to bat for you.

If you really want to try for the managerial ranks, it may be, as hinted above, that your best opportunity will be in another high-tech company. You've spent some years acquiring a broad range of experience, and you may think you have much to contribute in determining your company's strategic direction, yet you find that no one who counts will listen to you. The Bible says that a prophet is not without honor, save in his own country. That being the case, you may wish to change countries, metaphorically speaking. A man who had been with the same computer company for seven years, and whose last position was manager of relations with ISVs, reflected on this:

> I've accepted that it's nice to be at a company for a long time, but in the high-tech industry things change quickly and staying at any company for more than five years can be detrimental to one's career. Therefore I'm not going to commit myself [to his new employer] the same way I did here. I don't mind changing a company. If I stay with the next company for three years, I'll be happy.

Continuing Your Education

In the chapter on culture, we've already seen how the high-tech industry values training and retraining. Each quarter, you'll probably have the right to take one or two courses within or outside the company, depending on the training opportunities within the company. You should take advantage of these courses. You can pick up some in-depth technology; you can also learn lots of effective management techniques, such as how to manage in ways that respect gender and ethnic diversity, how to negotiate, and so forth. It's not necessary to add these classes to your resume, but they will make you a more mature and able business person.

Many high-tech employees are "too busy" to take advantage of training and educational opportunities. This is really too bad because it demonstrates mistaken priorities. You can always stay late or work weekends to complete a project, but the more you keep putting off that course, the more your career goes along without the advantages it can bring. So fix a date, keep it on your calendar, and just do it.

You also want to learn on an informal basis. A woman who works in customer service in a large hardware company said:

> Keep yourself informed by reading. Read the magazines, read the publications that are out there, and understand what your competitors are doing.

A marketing manager talked about survival in high tech in these terms:

> I would say that when I look at my peers, we all have a few things in common. We all basically follow what's going on in the NASDAQ stock record every day. So I know when Oracle's stock goes up, and I know what's happening with HP's stock, and IBM's. And all this following of NASDAQ goes under the heading of knowing what's going on in high tech. That also means I'm reading particular publications.
>
> And that information is important because it allows me to network, so I can actually talk shop with people and actually sound like I know what I'm doing. That's critical in [Silicon Valley]—knowing about the little interworkings and about company movements, and who acquired whom and who has hot technology. If you know something about a company or you know some people at a particular company and it's valuable to someone else, they can use you. And the favor will come back.
>
> I'm constantly on the Web, looking at things; I subscribe to magazines and Internet services send me things. And I do find that has helped me a lot.

SUMMARY

To thrive in high tech you must take the following steps:

- Grow your job
- Gain visibility through presentations
- Develop and use mentors
- Make lateral moves
- Continue your education

These five actions will ensure that you'll thrive in the high-tech environment and that you'll never be bored.

8

START-UPS

High-tech start-ups have entered American business lore and American folklore. Americans love stories of people starting from humble beginnings and then hitting it rich. We especially love them if we think we, too, have a good shot at following the same path to riches. High tech, with the phenomenal explosion of new companies, seems to offer such opportunity. Bill Hewlett and David Packard started Hewlett-Packard in a garage in Palo Alto, California, and built it into a giant company. Bill Gates, founder of Microsoft, dropped out of Harvard, but redeemed himself by becoming the richest man in the world. Apple Computer also got its start in a garage, with Steve Jobs and Steve Wozniak as the sole employees. And there are many similar stories of successful start-ups in the high-tech industry.

Start-ups are launched when someone has an idea and isn't content just to leave it at that. The idea could involve any aspect of the high-tech industry: hardware, software, consulting, publishing. Most people who launch a small high-tech company have spent some years working for large high-tech companies. They take their idea and share it with two or three other people; they massage and shape it, get an idea of the market once their idea is "commercialized," and quit their jobs to become entrepreneurs.

Of course, these entrepreneurs must have access to capital; sometimes initial capital comes from savings, relatives, and friends. The first round of financing is usually the "seed money," a small amount of money to research an idea and determine its marketability, whether it can be produced in a timely manner, and what resources it will take. This process may take as long as six months. Following this phase the venture capitalists ("VCs") may decide to fund the start-up, and that's when the significant cash begins to flow.

In the 1980s risk capital could actually be obtained from the banks; today that is rarely possible, as the banks demand security that the fledgling company does not have. Therefore start-ups often attempt to get funding from venture capital firms. In order to do so, a business plan is drawn up that shows how the company plans to develop, manufacture, and sell its products. The plan includes the history of the founders of the start-up, since the venture capital firm will depend on their technical know-how and business acumen for the success of the investment.

What are the chances of a start-up's obtaining venture capital funding? The head of a VC firm recently said, "In one year we looked at about seven hundred proposals. We decided to fund four of them. Two of these have gone sour already." But he went on to add that he was very satisfied with the two remaining start-ups; they were performing well, which could eventually translate into huge profits for the VC firm. These days, VCs tend to take companies public very early on. That way they can extract their profits and avoid the risk of the "early childhood diseases" of young high-tech companies. These diseases usually result from bad business decisions made as the start-up tries to go from being engineering-oriented to being market-oriented. There are many pitfalls along the way.

Another possible source of start-up funding is an "angel." An angel is someone who may advance money merely on the basis of a good idea—often without a prototype, or even a business plan. Angels tend to come in early and do very well indeed when the company is successfully taken public.

The wise start-up entrepreneur will have made an effort to sign on some impressive names to the board of directors; these people can provide introductions into the VC community, as well as credibility to the start-up.

After a couple of rounds of financing, the people who started the company generally own less than 50 percent of the shares—that is, they no longer control the company. It's now controlled by those who provided most of the financing.

When the venture capitalist firm decides to invest in a start-up, it usually wants and gets at least one seat on the board; the financing agreement may permit it to assume control of the board. This enables the VC firm to keep its finger on the pulse of the business as it develops. When the company is deemed ready to go public, the investment banks get involved. Such institutions have specialized sales forces that are looking for companies that might be ready to go public. A representative of such a bank told me:

> We are constantly out there looking at companies, and we make the initial decision whether this is interesting enough to bring in an analyst. We get involved at the top of the company. They usually have a corporate presentation they do for us, and we ask them a lot of questions. If it's interesting, I'll come back with my analyst in a week or so. Then we ask them for their financials. If we don't think a company is doing well, we would not take it public.

The investment bank looks at everything—the competition, how big the markets are, the company's strategy and how well it appears to be executing it, what technological advantages it has, and what is proprietary. *Proprietary*, in this context, means a significant market advantage that other people won't easily be able to match.

When the start-up goes public, the stock usually (but not always) rises, sometimes very sharply. After a few months, though, it's not unlikely that the stock will sink. An investment banker told me that he could pretty much guarantee that the stock of any company his firm took public would rise for about three months. After that, though, there were no guarantees.

WORKING IN A HIGH-TECH START-UP

Most people working in the high-tech industry believe that a start-up can only afford to take on very technical people. The company is under extreme pressure to bring out a successful product, and it is generally believed that there's simply no place for a nontechnical person. Though this may be true at the very beginning of such companies, this period doesn't last very long. Many start-ups have taken on nontechnical people when the total staff of the company was thirty people or less. In one start-up with eighty employees and another twenty contractors, half the employees were in engineering; another quarter were in sales, and the rest were in marketing and various support jobs. In the marketing and sales side there were people with no formal technical education. What they did have was experience. As one employee told me, "If you're a nontechnical person trying to break into the high-tech industry, a start-up is a very tough way to do it because there's no time to train people."

But nevertheless, I do know several people who've joined start-ups as nontechnical people. Some did it by being office managers—after all, someone has to deal with the landlord, see that the bills are paid, do record-keeping and filing, handle incoming phone calls, and arrange for the pizza parties. Others become Web page designers or writers. So it can be done. In the often frenetic atmosphere of a start-up, just being there and available for any assignment will soon establish you as a valuable member of the team.

What do you look for when trying to judge whether or not to take a position in a particular start-up? I spoke with a young woman who had given the matter a lot of thought before joining her present employer, a start-up with, when she signed up, about sixty people:

> I actually do have a list I made up for myself when I was doing this last search, based on the start-up I was in before that was good, and the first one I was in, that wasn't so good.
>
> On the financial side I look to see if they're a subsidiary of a larger company. That's a no, because it means they're not a true start-up. They won't have the freedom to decide to go public or to accept an offer from another company. They don't have that independence. And I want to see if their funders are legitimate VC sources, or if they're a "Chuck's Loan Shop"–type outfit.
>
> I look to see if it's too much [capital] or not enough. By that I mean that if you have too much, and you do eventually go public or are sold, then the value of the employees' stock is much lower. On the other hand, if you don't have enough, then you're constantly battling not having the right equipment, and you won't be able to make payroll. For me, I had a rough idea of how much operating expense a company of this [present employer] size ought to have, and I wanted to know how much money they had in the bank. That's something you won't learn until the final stages of an interview because it is fairly confidential information. They won't volunteer that information, but if you ask for it, you'll get it.

> The second thing I look for is the executive team. The CEO is the key role. I wanted someone who was very smart and had a good business picture, who wasn't afraid of making decisions, and who was secure and well respected in the industry. I look for the same things in the marketing, sales, and engineering teams. One of the things about [present employer] was that the main people had already worked together as a team and had an impressive record. I also had to be sure there was a strong engineering idea and a strong marketing team to support it because neither one is sufficient without the other. I'm not an engineer but I have friends who are, and I asked them to assess the engineering side by giving them pages from the Web, and white papers, and asking them to assess the technology.
>
> And finally there's something much less tangible. I wanted to come in and see if things were happening, people were enjoying it, people were staying on, whether people understood what the mission was.

What do such people do? One woman in her thirties who is with a growing software company of some three hundred people in the Boston area told me:

> When I started here I did everything. I did accounts payable and receivable, payroll, shipping, receiving, bought the lunches. They already had a janitor, so I didn't clean the bathrooms. I was the eighth person hired. Everybody else was a programmer or a marketing guy, and I did everything else.

The principal reason most people are attracted to start-ups is that their size means an informal and varied work life. An HR administrator in a small software company put it this way:

> When I was here four years ago, we had trouble getting people to do expense reports because it smacked of bureaucracy. The people that were here didn't like large companies, didn't like red tape. The small company atmosphere really attracted them. The fact that to do their job they probably had to do pieces of four other jobs attracted them.

And another start-up employee said:

> Part of the reason I like a small company is if you want to get something done, you go to the person who you think can do it and you talk to them. You don't have to go through channels and around barn doors and upstairs and downstairs.

So being in a start-up involves doing a lot of different things in an informal and fast-moving environment. And there's the added piquancy of risk—because needless to say, not all start-ups survive.

START-UP MORTALITY

Many start-ups do not make it past infancy, and for those who do, there lurk a host of childhood diseases for which no vaccine seems effective. People in high tech who think about start-ups—and that means almost everyone—spend the occasional lunch hour discussing why such and such a firm went under, or what the prospects are for good old Sam, who left two months ago to launch Softsmash, Inc. In retrospect, everyone can explain quite cogently why a healthy teenage company survived the perils of youth, but no one can predict with any confidence whether any given start-up will succeed, no matter how brilliant and hardworking the founders, how genial the idea, how well-funded the effort, and how eager the market. There are simply too many variables. Luck has to be with you as well as knowledge and skill.

The problems of nurturing a start-up through infancy to adolescence are many. As mentioned earlier, one of the main problems is that the founders, who are usually technical people, find it hard to make the transition to being driven by the market. They may think they understand the market, but their understanding may prove to be limited when they begin to encounter serious competition. The founders may understand intellectually that they have to get sales and marketing people aboard in order to survive, but finding the right fit and subordinating their creativity to the demands of a marketing department may be a step they can't successfully negotiate. The start-up is also under pressure to perform—that is, to produce and sell product—from the people who have provided the capital—the angels or the venture capitalists. Continued funding is always in question, as venture capitalists usually dole out the money in a series of partial transfers.

What is the mortality rate of high-tech start-ups? No one knows this for certain either, but probably no more than five out of a hundred will survive more than three years. Perhaps two of these will survive past the age of six. In the normal, non-high-tech world, being associated with a company that goes belly up sometimes carries a stigma; the company failed, so those who started it and who were employed by it must somehow also be failures. No such stigma is attached to the failure of a high-tech start-up because everyone knows, from the high-tech media to the VC firms, from the headhunters to the bartenders at high-tech watering holes, that a start-up is a risky business. People in the high-tech industry admire those who have an idea and are willing to back it all the way, even if that means all the way to the bankruptcy court. As a start-up employee put it:

> Everybody has their strong suit. You could do well starting a company, have the vision to start a company based on a product, and not have the skill set to grow it. And that doesn't mean you're a bad person. There are many more people who don't have the vision to start a company.

GETTING A JOB IN A START-UP

Whether a start-up survives through infancy does depend in part on how well it manages its initial growth. Often this calls for a skill set that the founders may not have. Founders usually concentrate on the technology and product development, but success and growth ultimately involve the support side of a start-up: sales, marketing—all the activities mentioned in chapter 3. Of course, a tiny company cannot afford all the people needed to do all these tasks, but neither can it afford to ignore them for very long. Start-up founders often have a very clear idea of the market for their first product, but this is usually only sufficient to get the start-up past the first year or so of production. That is often the entry point for nontechnical people. In Web-based start-up companies, nontechnical people have many opportunities to get in early—earlier than one could, for example, in a hardware start-up—because the Web-based company is dealing directly in communication, a skill many of us liberal arts types have in abundance.

For high-tech insiders and outsiders alike, the route into a start-up is almost always through a friend who gets involved and then brings you in. If the start-up has only a handful of employees, you'll have to meet them all and pass muster with every one. If it's been around for a year or two and has, say, fifty employees, you won't have to be voted in by everyone. If you are already employed in a high-tech company, it's particularly easy to find opportunities in start-ups, especially if you have friends in engineering disciplines. They'll know that such and such a person who just left has started a company or has joined up with so and so who started a firm last year. Word gets around.

There are two points at which you might get involved with a start-up. The first is at the very beginning. If you have no obligations to others (spouse, children, other dependents) and don't owe too much on your credit cards, you can try to be employee number ten at the start-up. Salary? Not much, but maybe enough to barely survive. Most of the compensation at this stage may be in stock options, which are convertible into stock after the company goes public, after some period of time. On the other hand, if the company is well funded by venture capital, perhaps you can get something resembling a decent salary.

The other point at which you might jump on board a start-up is a bit later, when there's already some history to the company and there appears to be a good chance that it will survive. Stock options may not be as plentiful at this stage, but there should still be some. In a typical case, a young man joined a software start-up as an hourly employee doing rather routine tasks. His total compensation the first year was only about $20,000, not including some overtime, but he received a vested interest in 1,500 shares of the company. For each of the next three years, provided he remains on the job, he will receive another 1,500 shares. After one year he received a raise of only 3 percent, but was granted an additional 1,000 shares of stock. Assuming the company survives and he stays with it and that the same rate of largesse is maintained and that the stock isn't diluted, he could end up with 10,000 valuable shares at the end of four years. He and his colleagues are hoping that the company will go public, that the value of the stock will soar, and that they can cash out at some point with a handsome gain.

BEYOND MONEY

People working in start-ups find that two of the most rewarding things about them are the fast pace and the responsibility employees shoulder from the beginning. A woman described her first days in a start-up as follows:

> It was so refreshing, and empowering, and daunting, also, because I realized I was the expert. And I didn't necessarily feel like an expert on day two!

Another thing start-up employees like is the variety. One of them told me:

> If I get bored working on one particular project, there are twenty other things that could use my attention. Whereas when I worked in a larger company, I had one specific area of responsibility, in a very narrow band. If I got bored with it, that didn't matter—that was it. So I love the variety here.

And what about the "get-rich-quick" aspect? I was told:

> I would say that's a big draw here. It's certainly, for me, one consideration because I can take more risks now with my life than I probably can five or ten years from now, because I'm just me. I rent; I'm very mobile; and I'm only responsible to myself. So I knew that if I were putting part of my long-term pay at risk, and it didn't work out well, it just affected me. And if it did work out, then it gives me a lot more freedom down the road to do things that I really like to do, like volunteer at a marine aquarium.

Working in a start-up gives you fantastic exposure to many aspects of the high-tech business world. Of course, there's a downside as well. The biggest negative is probably the pressure. An HR director observed:

> Every minute of every hour is jam-packed with deliverables and pressure. There's no downtime. It's very difficult to do anything proactive. It seems like, if you allowed it, you would have a reactive day all day long. And when I interview people, I try to have people understand from the beginning what a pressure cooker this is. I don't want people to come in, then realize it, and leave. So I try very hard not to hide the good, the bad, and the ugly about this place. I ask my interview teams to be very candid about the job, the hours, the pressure. We have futons in the hallway the engineers have used occasionally. They roll them into a conference room and take a nap. Someone asked me to hide them, and I said no way, I want the candidates to see that and know that it can be that nuts here. Sometimes we get candidates here who've only worked for large companies, and I think those hires are risky because it's so different here.

If it gets too rough, you can always come in from the cold into a large company; people do this all the time. You'll have your network in place, and you'll have experience. A middle-aged woman who became involved with high-tech start-ups after having spent some years as a housewife had this to say to those who are interested in start-ups:

> Don't give up. This is my third start-up. The first two didn't make it. The first one peeked at 130 employees and then died. The second wasn't ever going to be that big. We reached about twenty-two people and then it was clear we weren't going to make it. I knew the president of this company [current employer] from the second place. So when he became involved in a new start-up and they needed a support person, he called me.
>
> The mortality rate can be pretty high, but just because one start-up doesn't work, don't let it sour you on trying another if that's the type of culture and environment you're comfortable in. In the worst case, even if the company doesn't make it, you've had a job. Hopefully, they've been solvent enough so they can make payroll, but you've been making contacts all the time, and you're growing your skills unbelievably.

INTERNATIONAL WORK IN HIGH TECH

My high-tech career has been international; working from the United States, I have flown all over the world. A while back, I found myself commiserating with a friend of mine who was vice president of the Far East division of a hardware company. We were complaining to each other about the long overseas flights we had to make, the jet lag, the missed birthdays and other celebrations while we were on the road, and how wearying the whole thing was. We were silent for a while. Then he said, "Of course, if I weren't doing it, then someone else would be—and I couldn't stand that."

Working in the international end of high tech can be a lot of fun. It's all I've ever done since joining this industry, and though I can complain with the best of them about all the disadvantages and disruptions it brings to one's life, I can't conceive of doing anything else. A few weeks ago I was talking with the director of a university career center, and according to her, lots of young people are poised and eager to get involved in this side of the business. I get a fair number of requests for advice on how to "get into international."

You should be forewarned that international jobs in high tech are almost impossible to obtain as a first job. I'm going to cover the reasons for this in some detail later in the chapter. But you should not stress about securing such a position right away—they exist, and you can work your way into one after you get your first job.

What is it about international work that makes it such an inviting path to take? These are some reasons that I, and various colleagues, have come up with; I'm sure you can add a few of your own, if you've also got the travel bug:

- *Having more interesting job content because of the added challenges presented by the international environment.* You have to present your technical story, perform your sales pitch, or manage your staff in ways that may be quite different from those effective in this country. If you've got the right frame of mind, this adds an interesting dimension to your job and makes you as close to indispensable as an employee is ever likely to get. People who can function well in an international environment are rare finds.
- *The opportunity to get immersed in a different culture.* The degree to which this happens is generally up to you. Having sufficient time on a business trip to get out of the hotel or office and get into the local scene can be a challenge. Having a trip last through a weekend gives you the time to make that side trip, see the countryside, stay at an out-of-the-way inn, and so forth. Of course, if you're on an international assignment for a couple of years, you've got plenty of time at your disposal.
- *Using that foreign language you worked so hard to learn in high school and college.* Yes, it is true that most foreigners you'll come into contact with will speak English. It's also true that most of them will appreciate any effort you make at least to socialize with them in their native language. Language is perhaps the single most important element of a culture, and understanding how others express themselves is key to understanding how they think.
- *Extending the knowledge you acquired in an international area studies major in college.* If you live and do business in a foreign country, you will acquire knowledge and understanding that can surpass anything taught in a college course.
- *Having greater responsibility than you would have as a stay-at-home.* This is one of the most attractive attributes of international work. When you are overseas, you often *are* your company to the outside world. Do your business cards show that you are really only a miserable, lowly individual contributor? Explain to your manager that you are dealing with CEOs and vice presidents, and that you need more status, even if only on your cards. Often, however, though your foreign business associates start out judging your importance by your title, as you start producing results for them and building the relationship, they will start according you the respect you deserve.

- *As the sales representative for Widgetron, you may have much higher-level access in Bolivia or Beijing than you do in Boston or Boise.* You are not only a sales rep overseas, you are also the local source for all sorts of information—a convenient channel for inquiries of all sorts, relating to finance, engineering, quality, company policies, industry news, and so forth. This, in turn, means that you will be dealing with a wider range of people in your home office, whether in person or by phone or e-mail. This gets you noticed and builds your internal network.
- *Picking up lots of frequent flyer miles (so you can do more flying on your vacations).* Being able to fly overseas business class, with your significant other, for your vacation does make up somewhat for all those other trips you had to take without her or him.
- *Having an international assignment on your resume as a career booster.* Many companies want their rising stars to have spent some time on an overseas assignment. Overseas markets are immensely important for American business.
- *Having the great cuisines of the world available at the source.* This is an entirely appropriate reason for doing international work and one of which I approve heartily.

You'll note that the above listing mixes up work-related reasons with others that are mainly personal. Whether you're living overseas or spending a lot of time overseas on trips from the United States, being able to function effectively in a foreign environment is actually part of your work assignment. Therefore, I believe that you find yourself "on the job" for a much greater period of time each day than if you were working in this country. The distinction between personal and business rewards tends to get blurred.

GENERAL CONSIDERATIONS

The United States is clearly the world leader in high-tech innovation and commercialization. That being the case, American high-tech firms carry out a lot of business with governments, businesses, and various organizations in foreign countries. Perhaps the only high-tech enterprises that don't engage in some kind of international transactions are very small, very young companies that don't have the know-how or simply haven't had the time or the resources to get started. But international business cannot be ignored for very long. A mature company may get up to 50 percent or more of its revenues from overseas sales, and it's very common to have overseas software development or hardware manufacturing operations as well.

Basically there are two types of international work in the high-tech industry. In the first, the employee is based in the United States, but the job requires frequent travel overseas. The job might be in channels management, sales, or business development.

Why not just put these persons overseas? Sometimes this does happen, but it may also make sense to have the channels manager position located in headquarters because a lot of what this person does require daily interaction with functions located there—marketing, finance, product management, public relations, and so forth.

Similarly, it may make sense to have the salesperson located in the United States and making periodic trips overseas, if this person is of U.S. nationality and not a "local hire," because, as we'll see, posting someone overseas in an "expat" position can cost hundreds of thousands of dollars each year. Such a person may still have to make periodic trips back to the United States for product training and sales and other meetings. Also, it's generally regarded as undesirable to have a salesperson in a foreign country, alone, without the moral and business support of a local staff. Such people tend to get lonely; they may also be superb candidates for poaching by the competition.

The business development person, like the channels manager, usually has to work intensively with headquarters people, so it makes sense to keep her at home, even though extensive travel may be required.

The second type of international work involves an overseas posting. These jobs are also found in a variety of functional areas—more, if the company is on a phase of "moving functions out to the field," less if the company is retrenching and pulling functions back to headquarters. If getting into international work and living overseas is your prime objective, you ought to find out which stage your particular target company is in—expansion or contraction. Be assured, though, that these things go in cycles and that, even in a company that is pulling back, there will be opportunity for overseas work at some point.

When researching this information, if possible, do so with great circumspection, or from a source that will have nothing to do with your hiring process—an annual report, for example. Generally speaking, it is not a good idea to let a recruiter or hiring manager, or anyone in your interview cycle, in on your fondest dream: to be sipping coffee on a warm spring afternoon on the Via Veneto, celebrating your latest billion-lira sale as Widgetron's Southern Europe sales manager (unless, of course, you are specifically applying for this position). More on this later.

There's a third way to get overseas with high tech, if you only want to travel infrequently. There are many positions where a bit of travel is part of the job, including marketing, legal, finance, sales support, contract negotiations, credit management, and account management positions in a high-tech company's international division (if you have such a position in a corporate headquarters division, your chances of international travel are less). This chapter does not address this sort of position, but these days almost any of the jobs listed in chapter 3 could involve some international travel.

INTERNATIONAL WORK BASED IN THE UNITED STATES

I've already mentioned three types of jobs that can primarily involve international work: sales, channels management, and business development. Unless you've already done work like this for another high-tech company, though, it's very unlikely you'll get a chance at these sorts of positions as a newcomer.

My favorite category of jobs that might get you a chance to travel internationally in your first high-tech position is that of customer service. This is a broad category (see chapter 3), and since it doesn't seem very interesting at first glance, many people seeking jobs overlook it.

Remember this: Any work that takes you close to the international customer (whether *customer* means the end user of the product or a sales channel) is good work. Customer service positions tend to survive longer in times of economic downturn (no industry can ignore its customers and survive). More important to you, however, is the fact that these sorts of customer service jobs don't tend to attract people who are as obsessed with working internationally as you are. That gives you more leeway to move into the international side of things.

Since there are so many variations in customer service work, it's very hard to give a comprehensive description of them, but here's an example of how a friend of mine parlayed a customer service position into some pretty decent trips.

She got an entry-level job as a customer service representative (CSR), dealing with distribution channels in Europe. International customer service often involves dealing with customers across many time zones. When it is 9:00 A.M. in California, it's 5:00 P.M. in Europe. In order to have any kind of meaningful phone contact, therefore, either the U.S.-based customer service rep has to get started very early, or the European customer has to work late (indeed, in Europe people tend to stay around until 6 or 7 P.M. or later). In my friend's case, she eagerly took the CSR job that required her to begin her day in the office at 6:00 A.M. No one else was terribly interested in doing this.

After a few months she became indispensable to the European sales staff of her company. She never let things slide and always got back to the customer, even to say that she had no answer yet to the particular problem but would get back to them soon. She got herself invited to a European sales meeting by offering to make a presentation there on the U.S. order operations process. The European sales manager told her U.S. boss that he needed her to explain the process, so her boss let her go.

She now travels to Europe three times a year, varying her stopovers between London, Paris, and Munich. She could have a job in Europe if she wanted to move there as a local hire, but she prefers staying in the United States for the time being.

Many high-tech companies haven't really thought out all the support implications of their international business. I'm a firm believer in spotting these opportunities and then either growing your job to encompass them or just volunteering to do them until the international sales folks realize they can't do without you.

Actually, the process I've described above can apply to many other parts of the high-tech business, such as public relations, human resources, employee communications, and so on. The trick is to spot the opportunity and then make it happen. If your knowledge of American business is limited to the calcified giants of the automotive world or the cereal makers or the insurance industry, this must seem like fantasy to you. But high tech is different. It's fast-moving, exploding in size, and has vast international market potential. Once you are an insider, willingness to work counts for a lot.

OVERSEAS ASSIGNMENTS

What is the position of the high-tech industry today, with respect to sending Americans overseas? Actually, pretty good, compared to the situation during the massive downsizing of the early 1990s. At that time an international HR manager told me:

> [In the past] I felt that it was essential that our American employees [get overseas]. But that was a few years ago. Now we're in more competitive times and everyone is questioning all areas of cost, what the return on investment is in these expatriate assignments. It's very difficult to quantify the return. So now people are questioning every assignment and suggesting it's not a good thing to do.

Since then, however, things have changed a lot. The huge popularity of the Internet and the multimedia and other technologies it has spawned have resulted in an almost incredible expansion of the high-tech industry over the past three years or so. Successful companies—including some that hadn't gone public three years ago—are now busily expanding into overseas markets. A young software company recruited a friend of mine to be based in London and to build an international operation from scratch.

Successful larger companies are increasing their international presence as well. There seems to be less reluctance to make the sizable investment required to put an employee overseas. For one thing, at this writing, the high-tech sector is doing very well indeed; there is simply more money available. Still, finance managers' memories tend to be longer than those of sales managers; the investment will have to be justified. And, of course, sending an American often will have to be justified versus hiring a local person to do the job.

To give you an understanding of the costs involved, here are some figures. It costs a company between $300,000 and $500,000 a year to send an "expat" overseas, depending on the country. If these figures leave you incredulous, consider that the rent for a first-class apartment in Hong Kong (as of the end of 1997) runs from $10,000 to $12,000 per month. That's U.S., not Hong Kong, dollars. It is necessary to pay between one to two years' rent in advance, and of course, that does not include any furnishings. Furnishings must be purchased locally or shipped from the United States. If they are shipped from the United States, the cost will be immense for the usual weight allowance of 6,000 pounds, which is not enough to furnish a whole apartment. Of course, this does not include major appliances, which are usually purchased locally because of voltage differences. If you're an expat, the company will pick up the bill for these appliances.

Other expat costs usually include: private schools for children (a necessity in some places, including Hong Kong); managing the rental of the U.S. home; loss on selling cars (or storage of cars); tax advisers' fees; cost-of-living adjustments; hardship post adjustments; hotel, transportation, and restaurant costs while searching for a suitable dwelling (and while waiting for furnishings to arrive); the cost of an automobile in the foreign country (often two to three times the cost of an equivalent vehicle in the United States); air fares to and from the assignment, as well as for home leaves and vacations; air fares for college-age children to and from the overseas post; medical evacuation insurance; storage of household goods while on assignment; household staff in the country of assignment (sometimes a necessity); and so forth.

Sobering, isn't it? In addition to your salary, the company will be forking out a lot more cash with your name on it. You can imagine how carefully they'll scrutinize you before they send you. So you have to position yourself very carefully, in order to pull off an overseas posting.

WHAT HIGH-TECH FIRMS DO OVERSEAS

High-tech firms do three kinds of things at overseas locations. First, they have hardware manufacturing operations, often for reasons of cost, better proximity to markets, or both. In many countries it's possible to hire a line worker for a manufacturing plant for a daily wage that is less than what a

U.S. worker makes in an hour. This is why many manufacturing jobs are leaving the United States—and other costly countries, such as Japan—for "offshore" locations. These jobs aren't always in the low-skills category. Countries such as Korea, Malaysia, and Mexico are home to sophisticated manufacturing and assembly operations calling for highly skilled workers. China and Taiwan are rapidly becoming known as locations where high-tech equipment can easily be produced to a world-class standard of quality. Many other countries, such as the Czech Republic and the Philippines, are taking steps to invite high-tech firms to establish manufacturing operations.

High-tech companies have also established software development joint ventures with overseas companies. India, in particular, is known throughout the high-tech world as a source of excellence in software development. In the United States a reasonably good software developer costs a company about $80,000 to $90,000 a year, exclusive of benefits. In India a developer of the same quality costs about $9,000 a year. Considering that software can be sent electronically from India to the United States virtually instantly, and for almost no cost, it is easy to see why a very large proportion of the software development industry may move offshore from the United States to India in coming years. This movement is already well under way and has been facilitated by the Indian government's lowering of barriers to investment and repatriation of profits and capital by foreign firms. Similarly, software development for U.S. companies is taking place in Russia, including in Siberia.

The third type of activity that American high-tech firms carry out overseas is perhaps of more importance to nontechnical people. It involves the sales and marketing of products.

Sales

There are certainly lots of reasons for having a local sales force in an overseas market. Because knowledge of the market, the competition, and the local language and customs are important, as is having a network of local contacts already in place, sales representatives are usually, but not always, local nationals. But Americans from the home office may be placed overseas in sales positions for any or all of the following reasons:

- They know the products and have experience in how to sell them, and for competitive reasons the company needs to move rapidly into the market.
- They have special knowledge of an important market segment, such as banking, and it will take time to train a local salesperson or locate one with equivalent knowledge.
- There is already a large local sales force, and one or more "role models" are needed to demonstrate how the company does business.
- The company wishes to give a sales rep exposure to international business before promoting him or her to a management position.
- The local sales force is young and lacks credibility in the marketplace; a representative from headquarters is needed to make joint sales calls to build credibility.
- The competition uses Americans to project how important it considers the market; the company has to match the competition.
- The market has hitherto been serviced from the United States, and the sales representative is now being posted overseas until the transition to a local sales force can be made.
- It's more prestigious, in the customer's eyes, to be called on by a foreigner than by the local representative.
- The overseas sales office is a regional office servicing many countries, and local nationals are not welcome in some of these countries.

Marketing

Most high-tech companies have both corporate marketing and field marketing operations. The corporate marketing people have a lot to do with the public image of the company and developing the "party line" (i.e., what sales reps and others should say) with respect to new products or concerning the various product-related problems that occur every now and then. The field marketing folks, on the other hand, are the ones that most directly support the sales force.

It makes good sense to have the planning of field marketing activities done in the market they address, so many high-tech companies will have a number of marketing positions in foreign countries. These deal with advertising, trade shows, local materials of various kinds, such as product description sheets, and so forth.

Other high-tech companies try to develop international advertising, public relations, and promotional programs from within the United States. The idea seems to be that one can save money by piggybacking on programs developed for the U.S. market. An amusing example of this was Sun Microsystems's ill-fated attempt to have as its mascot a huge dog named "Network." The idea was apparently developed in the United States, but never took hold overseas, where poor Network became the focus of much abuse by company employees because of the reportedly fantastic amount of money expended on the canine image, at a time when money seemed to be lacking for more traditional marketing activities.

In fact, what often happens in high tech is that the United States sucks up 80 percent or more of the mindshare and resources of field marketing. It's often very hard for overseas sales to get the marketing support it needs when, and in the form and quantity that, this support is needed. Here's a great example that also indicates the desirability of having a person from the U.S. filling an overseas slot.

A French friend of mine was the sales manager for France in a U.S. software company. He wanted to have his company's products featured in an important show that was coming up in six months' time. Registration for participation in this show was almost closed. He needed about $50,000 to get booth space and various collateral materials and other support for the show. But his local field marketing people had no independent resources; they depended on the largesse of the American field marketing organization, and they didn't have the clout to jar loose any funding because they were local employees, with no personal connections to the people that counted.

My friend, looking for money for the show, e-mailed field marketing back in the States. No reply. Another e-mail; still no response. He telephoned and reached voicemail. Frantic, and with two days remaining to sign up for the show, he telephoned a personal friend in the U.S. headquarters. This man worked down the hall from the field marketing people at headquarters, and my French friend asked him to walk over and personally persuade the HQ field marketing person to get on the phone.

Finally, with the connection made, my friend explained that he needed the commitment of $50,000 immediately. The answer from the field marketing person was: "Oh, we had a meeting a couple of weeks ago and decided we weren't going to participate in any more trade shows."

This decision was made based on what the trade show schedules and projected demands had been in the United States. The overseas offices hadn't been consulted—no one had thought to do it—so the money wasn't available. If there had been an expat doing field marketing, she might have had the connections to resolve the problem and also could have ensured that her local counterpart developed similar connections.

If the above example seems preposterous—after all, systems ought to work better than that—let me assure you that this is not an unusual situation. Remember that high tech is extremely fast-paced, and often decisions are made on the fly in the United States, without anyone thinking to consult the international offices. Personal connections can help you avoid a lot of grief.

OTHER POSITIONS

Many high-tech companies have regional offices overseas, and these regional offices may have positions in finance, legal, operations, and human resources. The role of the regional office is to coordinate business in countries where the company may have distributors and other resellers, or even a direct sales force, manufacturing operation, and research and development laboratories. It's very possible to secure an assignment in one of these areas, but you're unlikely to as a new entrant in the high-tech industry. Your chances are much better after you've won your spurs by putting in some time in a similar position in the United States.

Newcomers

I've been in the international area my whole working life, so believe me when I say I sympathize with people who want to live and work overseas. The truth is, though, that it is not realistic to be a newcomer to the high-tech industry and expect to be in the running for an overseas assignment. Over the years I've been approached by U.S. foreign service officers and others with substantial experience, people who speak one or more foreign languages well, and even some who understand the industry and its international business issues as well as is possible from an outsider's perspective. They've wanted to know what the opportunities are for them to work overseas for high tech. I tell them what an HR manager told me in an interview:

> U.S. companies in our industry are unlikely to recruit somebody from outside to go overseas. They want to send their tried and trusted lieutenants. Being four or five years in one company first of all and having some high-level performance definitely puts you on that platform. . . . These are jobs for people who have very strong communications and interpersonal skills, very strong skills of persuasion, good negotiation skills, and who are trusted by headquarters management. . . . Having technical skills would be a plus, but a person with technical skills who didn't have the others would fail in the international environment. If you have those [nontechnical] skills, there are a number of roles you can go after; if you have an EE [electrical engineering] degree, there are maybe one or two.

So there is good news and bad news. The good news is that nontechnical people actually have a better shot at getting an overseas posting than technical people. The bad news is that you can't just sail into such an assignment; you have to have a track record. Here's an actual case history of how a young, ambitious woman made this happen for her. She started by getting a lower-level job in an international division of a hardware group:

> I started out as an admin, but after a few months we got a new VP of marketing. He came and asked me what I was doing in that position, and he promoted me to being the program manager of the ISV relationship program. I asked for this because I could see that soon the [availability] of applications was going to be the most important part of the sale. I was to take the ISVs' software that ran on our machines and get it localized. That means arranging different levels of support for the software in different countries, and also, certain software tools needed to be put into the local language. . . .

> I grew my job, broadened it to include how you signed up a vendor. How do you send them a contract, get them on board? Then I started teaching the managers in the field to take that responsibility as a local responsibility, overseas. Then I could take a step back and get more strategic, decide who were the vendors who were bringing us the most business. . . .
>
> I had decided very early on that I wanted to join a company in some kind of international capacity; I wanted the company basically to send me around the world on their books. . . . It was a matter of climbing the ladder and moving up until you became more of a senior manager owning the charter. And from there, in order to understand truly what the business was about, I had to get out of headquarters and move overseas, close to the customers.

She took an assignment in Australia as a business development manager. It took her a little less than five years to go from being an admin to being a "senior manager owning the charter." She had grown her job and won the respect and trust of the vice president of the international division, so she got her wish.

But why, you may ask, should I have to wait five years to get a great international assignment like that? And you're very likely to raise this question if you've recently been awarded a master's degree in, say, Australasian area studies. These area studies programs are marvelous, and it would be great if every American businessperson working internationally had earned one. Many people I've spoken with who have wanted to work in the international side of high tech seemed to feel that their degree gave them an advantage over others. This notion is, of course, reinforced by the educational institutions, who need to sell these programs to potential students. The problem is that just having this as your background doesn't win you the position you covet. Here's a story that illustrates this point.

Several years ago I worked for a high-tech company that, like every other, really wanted to boost its business in Japan. I happened to meet a young woman who was just out of a prominent Midwestern university. She had a bachelor's degree in computer science, and she had spent six months in Japan in her high school days and had learned Japanese, in which she had a college minor. She wanted desperately to get back to Japan; secondarily, she wanted a job. I passed her resume and cover letter to a friend who, I thought, would be interested in interviewing her, especially since we were expanding operations in Tokyo, and there would probably be some headcount soon. To my surprise, my friend did not seem in the least interested. He put the young woman's resume on a pile of folders on the shelf and, pointing to the stack, uttered the memorable phrase, "They're a dime a dozen."

What? Japanese-speaking, excellent technical education, and a woman to boot (we needed to do better in the EEO department), and she was not worth even an interview? Did anyone counsel this young woman during her college days that even an insider would be unable to get her a lousy fifteen minutes with a hiring manager?

Several years later, looking back at this incident, and with the benefit of having witnessed other similar situations, I can see what went wrong. In the first place, she had no work experience. If she had taken any sort of job with the company in the United States to start out, she wouldn't have had this disadvantage. In essence she was asking the company to invest a good deal of money in her and to take a substantial risk by sending her to Japan. What if she had got to Japan, decided she didn't like her work situation, and had simply resigned? That wouldn't make the hiring manager look very good.

Also, in her cover letter, she came across as very eager—much too eager. The great French statesman Talleyrand once advised his ambassadors: "*Surtout, messieurs, point de zèle,*" which is probably best translated today by the phrase "stay cool." That's excellent advice for anyone who wants to get into an international in high tech. Hiring managers will never hire you for an overseas job just because getting overseas is what you want most in the whole world. They are not in the business of fulfilling wishes. Therefore, it pays to play down your desire and play up your desire to do the work and the skills you have that will enable you to be successful.

When the young woman appliant specified Japan and enthusiastically emphasized her experience in that country, she unintentionally (but truthfully) sent the message that getting back to Japan was of primary importance, and the job itself came in second. She should have tried to give just the opposite impression.

If you have a good grounding in some international area and you speak a language or two, you're better advised to investigate whether the company has some U.S.-based support activity where such knowledge could be useful. When you're talking high tech and a command of the Japanese language, for example, that makes a lot of sense. High-tech companies do a lot of business with Japanese customers, and there are lots of contacts with those customers in the United States, through customer visits to headquarters.

If the company doesn't have such sales support activity, you might be able to create it—because undoubtedly the need is there—once you're safely hired. Don't suggest doing so during an interview for some other position. The hiring manager is not interested in your views of how the company's international business practices can be made more effective. Get on board first.

Your foreign language skills and area knowledge will not escape the attention of the hiring manager (provided you include them in your resume), even if she doesn't allude to them in your discussions. Careful! Sometimes hiring managers will bring them up. They may even ask you why you haven't chosen to capitalize on this aspect of your education by heading straight for an international job—perhaps with a bank or the U.S. government. This isn't a trick question; it merely reflects what the hiring manager thinks she would do, given the same background. If this should happen to you during an interview, for heaven's sake, don't start agreeing with the manager that maybe she's right and you *should* redirect your job search elsewhere. Keep your eyes on the prize, which is the position you are interviewing for. By all means acknowledge your interest in international opportunities, but state that for the next few years your strong interest is in getting a good grounding in the basics of the business (or in whatever specific opportunity is offered by the position you're discussing). She will be impressed.

HOW TO GET INTO INTERNATIONAL

It isn't easy to get into the international side of high tech, but you can do it. If you're coming from outside the company, the most important thing is to get a job—any job—inside the company. Remember that hiring managers will generally not take a chance on sending a new employee overseas. The cost and the risk are too great. The ideal situation is for you to get a domestic job within the international division, if there is one, or at least in a place that keeps you fairly close to the international folks. The next step is to learn who is who in the international setup.

Perhaps the most influential person for getting you the job you want overseas is the overseas sales manager. Successful sales managers generally get what they want from management, assuming that the headcount and req issues have been solved, so if the sales manager wants you as the in-country marketing or sales or HR person, you are virtually a shoe-in for the job.

OK, so how do you get this to happen? You want to position yourself as a valuable member of the sales manager's team—even when you're not officially on the team. How do you do that? Let me give you an example.

Let's say you want to get yourself to Argentina. Business is booming there; it's a good career move; you speak Spanish; you like to eat steak; and you'd like to do some skiing in Bariloche. The first thing you've got to do is meet the sales manager. That'll be fairly easy, as he will come to headquarters from time to time for meetings. You introduce yourself and tell him how much you admire Argentina, and find out what kinds of support from headquarters he needs but isn't getting. Don't worry—there will be plenty.

Then you provide as much assistance to him and his team as you can, given the demands of your regular job. For example:

- You get data sheets, pricing information, and PR releases to him as soon as they are available.
- You offer to act as a host for visiting customers from Argentina.
- You offer to serve as an informal single point of contact for him and his sales team. That is, you will always respond instantly to e-mail, phone, or fax pleas for assistance. (This can be invaluable in companies where formal communication channels for overseas operations aren't very efficient—that is to say, in most companies.)
- You have lunch or dinner with him, or members of his team, when they are visiting headquarters. An invitation to your home for drinks or dinner can work wonders.

In short, you build a reputation with him and his sales reps for being alert to their issues and able to help solve them.

If this sounds simplistic, believe me, it can work very well, as long as other pieces of the picture are in place. For openers, you are going to find out from the sales manager, who is now your friend, what positions are going to be opening up in the next several months. You are likely to know about them before anyone else does. Once the relationship is solid, you can make it very clear that you are interested in getting the job.

To make this happen, the following also must happen:

1. You must have an excellent track record with your present boss. Your sales manager friend will, at some point, ask your boss how you are doing. If the answer is "she's a top performer," your sales manager friend is going to realize that in getting you on board, he's getting someone special.

2. You will need support from human resources. Ideally you've cultivated the appropriate person or persons, and they will endorse you for the job.

3. If it's a sales or marketing position, you have to speak the language, and pretty well.

4. You must have acquired knowledge of what the business issues are with respect to the country or territory.

5. Ideally you've dealt with one or more important customers from the country, and they've said nice things about you.

6. You should be aware of living conditions in the country, and the tradeoffs you are willing to make in terms of giving up parts of an expat package. It is extremely rare nowadays that anyone is sent overseas with a full expat package—all the goodies mentioned earlier in this chapter. In fact, in most of the high-tech industry today, a full package is given only to people who really do not necessarily want to work overseas, but who have some special skills and talents that are required for a limited period of time. In order to get these employees to go, the company offers a full plate of inducements. The company has made the calculation that it's worth the high cost to get this person overseas for a year or two. You might have to go as a "local hire."

Of course, what I've outlined above is only an outline of the process; it will be necessary to make adjustments for the particular situation. In general the principals don't vary: Get into the company anyway you can, and then bring your resources to bear on getting your international job.

This even holds true when you've been working for another company in an overseas job, and you're now after a similar position in Widgetron. I think that the best you can hope for, to start off, is a job based in the United States that takes you traveling. Not that there's anything wrong with that; it's the kind of job many people are dying to get.

It's easier to use the steps outlined above to get a posting to a "less desirable" country than, for example, France. A friend of mine did this with the manager of the South Asia office of his company. The manager said he was looking for someone to do marketing in India. My friend let him know he would be very interested in this position. It was hard for the Singapore manager to find anyone to take this job, probably because India seems to have acquired a reputation as a difficult place to live (it is actually a fabulous place to live). So my friend got the job. He went as a local hire, but he did get a rental allowance and a car allowance, and he's living comfortably in Bangalore.

Keep in mind that, as we've seen, living overseas, even in a developing country, is more expensive than living in the United States. Though the U.S. government gives a tax break to people sent overseas by their companies, it's rare nowadays that the employee himself can benefit directly from it; companies generally insist that the employee pay the equivalent of his U.S. taxes into a company fund, which is used to offset foreign tax liabilities of all expat employees. And since we're on the nasty subject of taxes, some state tax authorities continue to insist that employees sent overseas owe them taxes, even if the employee has retained no connection with the state (e.g., no real property). So generally there is no financial advantage to an overseas posting with respect to taxes.

High-tech companies may offer to send you overseas as a local employee; they give you a salary and a round-trip ticket if you're lucky, and a one-way ticket if you're not. I have another friend who's a U.S. citizen, born in Hungary. He accepted such an assignment to Hungary because he wanted to be closer to his parents and also because the salary offered is very high by Hungarian standards. But he's a bachelor. Suppose you're married with children. Sometimes companies will pay for tickets for your family, and sometimes not. Costs of private schools, home leaves, a local automobile and driver, and so forth are out, for the most part. In some cases you might get assistance with locating a place to live, and even some help with the rent (usually this is worked through the local hiring manager at the overseas location). The company will assist you to obtain a local work permit, usually valid for two years (and very difficult to get renewed), but only one person in the family benefits from this permit—you. Your spouse may not legally work. Your children will have to go to local schools, and you'll have to rent or purchase local transportation yourself. You'll get no assistance in renting out your home in the United States.

This kind of arrangement can put a lot of stress on a relationship. In fact, it can end a relationship. On the other hand, if you are single, as is my friend in India, relatively young, and ambitious, and you really have the desire to go, it's well worth pursuing. One benefit is that in the foreign business community, say, in Hungary or South Africa, you are almost certainly going to occupy a more exalted position than had you stayed at home. Networking is pretty easy, and it's not unthinkable that, as your assignment draws to an end, you might make a deal with another company, either to stay on with them or get hired back in the States.

Having an international stint on your resume is very desirable in the high-tech industry. The paradox is that this important experience will most likely pay off for you in another company, not the one that sent you overseas. We'll take a look at the reasons for this in a few moments.

Here's how one person worked his deal out for a two-year assignment in Asia. Note that in this case, the employee ended up negotiating directly with the overseas operation, bypassing his U.S. HR people:

> My package is the most convoluted you'll ever come across. I don't know what label to give myself. The company didn't give me the option to go over as a traditional expat. An expat to me is someone who gets paid in U.S. dollars, who goes overseas, who has money sent to them, and who gets a whole lot of other benefits. . . .
>
> The position I moved into had been open for nine months. They had interviewed locally and couldn't find anybody, so it

was very much in the company's interest to send over a head office person who was in charge of that area at home, and to plug them into the country. Because I knew all the contacts at the head office, I knew how to jump-start the program very quickly.

[The U.S. headquarters] did ask me to go, but I also wanted to go. That turned into a big issue, because if the company sends you overseas, you must go as an expat. If you want to go, then they have leeway to kind of pull back elements of the package and say you're going because you asked to. It was a very lousy game they were playing. In the end I went over and decided to be paid as a local in local currency because the salary was about $40,000 more a year than what the U.S. [HR] people were offering me. I went through my own spreadsheet and realized I was actually better off with the local package. Yes, I had to pay local taxes, but I got a credit against U.S. taxes. I did negotiate assistance with housing, and I had a company car in the end. So I got a couple of perks that were important.

WORKING IN THE OVERSEAS OFFICE

The high-tech industry is fast-paced, hard-driving, and values "pushiness"—in the United States. Needless to say, not all foreign cultures take kindly to these characteristics, and when U.S. high-tech managers or individual contributors are assigned overseas, or assign themselves overseas, there can be some rough periods before smooth working relationships with nationals of the country of posting are established. We tend to take our culture with us, and sometimes we're quite unaware of how our behavior may affect our colleagues in other countries. A rather minor example of this occurred in my presence recently, when an American gave a presentation on certain international business issues in his company's Asia regional office. The slides had the headings "United States," "Europe," "Japan," and "ROW." *ROW* stands for "rest of the world" and is often used in high tech to designate countries that aren't in one of the first three areas. During the presentation one of the Asian employees leaned over to me and whispered, "I just hate that!" To him, *ROW* had a connotation of *afterthought*, or perhaps *insignificant*.

Here's an interesting perspective on the differences in culture, as seen by the wife of an executive who worked hard to get an overseas posting and finally succeeded:

> He [the husband] wanted to do this; this was definitely something he wanted to do. . . . I don't think he really understood how different each aspect of life is. . . . I think a lot of people have this image of how romantic going abroad is . . . and how important they are going to feel because they're going to be an American in Brazil or Mexico or wherever. . . . I don't think they understand how difficult it is sometimes to deal with regular business issues that you deal with every day in the U.S.
>
> Americans tend to think, especially if they're working for an American company, that company procedures are such and such in New York or in San Francisco or Los Angeles. [But] those company procedures may be completely different in the context of a different culture. . . . He would come home and say they don't do things this way or that way, in the way the home office wants them done. . . . I don't think we ever reached the point where we were involved in the culture because it's very easy to isolate yourself and become friends with other foreigners.

There are, however, much worse examples. Here's the story of a very successful manager in the United States, who went overseas to run a sales operation (the story is recounted by one of his employees):

> When Tony got over there, he was in typical Tony mode, very aggressive, fast-moving, loud and hard-charging, like most top managers at [company] are. And everybody turned off to him. And within two or three weeks he had absolutely no support from anyone. He realized after a while that you can't just come into a different culture and start pushing things around the way you want them to go.

Tony woke up too late; most of his best local nationals—managers and individual contributors—left for other employment in the first three months of his tenure.

Even though you may consider yourself to be culturally sensitive, too may find yourself in a delicate position when it comes to fitting into an overseas operation. A woman who took over a marketing position in Europe told me:

> You have to prove yourself to them—not that they're skeptical, they just want to understand what you're all about. And why did we have to bring someone in from outside the country when there are plenty of capable managers in the country? So they're looking to see you deliver some unique value added that is not available in the local marketplace. . . . For the first four to eight weeks I just walked around on eggshells.

REENTRY

The most traumatic part of working overseas in the high-tech industry could well be your reentry, or attempted reentry, into the U.S. company. If you went over as a local hire, you might run the risk of not getting a permanent assignment back home. The only way you'll get one is through your network, and it's hard—but not impossible—to create or maintain a network from overseas.

Even if you're a permanent employee of the company, you're going to have some problems. Few if any high-tech companies have succeeded in setting up a workable system to facilitate the reentry process, and as a consequence, the company's investment in expat employees is often wasted. Both expat and local hire employees are sometimes forced to leave their company because there is no job for them when they return. If you accept an overseas assignment, you should be prepared for this and take preventive action.

Companies sometimes have the trappings of a system to help returning employees find employment. When the employee first leaves for the assignment, he or she may be assigned a "sponsor" or mentor in headquarters, whose job it is to keep in touch with the employee and assist with relocation into a domestic assignment. This type of system usually doesn't work. Usually the mentor is in a different job, division, or location by the time the employee is ready to return. In any case, other employees who did not take overseas posts have already targeted the choice positions and have been working for months to get them; the returning employee is a latecomer, having missed out on two or three years of contact and political developments. It is extraordinary that, having selected employees for costly overseas assignments because of their special and even unique talents, the high-tech company does not go to extraordinary lengths to retain them after their assignments are over.

Most high-tech companies are still rather unsophisticated when it comes to tackling overseas markets. Customers in foreign countries like to do business with people they know and trust. In some parts of the world it can take two years to build such a relationship; at that point the American employee is either yanked back home or told that he or she can stay on, but only as a local hire. In other words, the employee must give up the expat package. Many European companies approach their overseas markets quite differently. They have what amounts to a minidiplomatic service, and employees serve in it for twenty years or more, moving from one post to another, but always within a given region. In Southeast Asia, for example, a British company might post someone in Singapore for three or four years, then move the person to Malaysia for a similar period. This might be followed by a posting to Indonesia, then to Manila, then to Hong Kong, and finally back to Singapore. Such employees become true experts in foreign markets, get to know important government officials on a personal basis, speak the local languages, and generally integrate better into the assignment. European expats, therefore, don't have any problem with repatriation and reintegration into the company because from the start they are in it for the long haul.

Most American high-tech managers don't agree with the European approach. They point to the success they've had in selling overseas as proof that the American system works. Fortunately for them, U.S. technical leadership in the high-tech industry has overcome shortcomings in their approach to sales and marketing. Channels and customers may not always be overwhelmingly happy with the service they receive, but they put up with it to get the products. However, when the technology of a U.S. company doesn't deliver the goods, foreign competition often wins the day.

The problems faced by returning U.S. expats are illustrated vividly in the following remarks by an individual contributor who has put in two years in Australia for a hardware company:

> [The U.S. company] wasn't interested in me because they were trying to reduce headcount when I returned. I received almost no attention and no assistance as my visa began to wind down. When I finally started to make my own arrangements to ship my stuff back, and the company woke up and realized they were in a poor position and could be sued.
>
> The company then sent me layoff papers before I even got back to the United States. So I cut a deal and said: "Contractually you are obligated to find me a job, so give me two months to look around and if I don't find anything, then you give me a severance package and I'll leave quite happily." So that's what happened.

It's amazing how many returning expats speak of having thought of suing or threatening to sue the companies that sent them overseas. That's because of the breakdown of the reentry process; no one wants to take responsibility, no one has a suitable position for the returning employee, and everyone wishes the problem would go somewhere else.

This sort of problem can even affect those who are very well entrenched in a company. Here's the story of a senior manager who had been with his company almost since its founding, who was very friendly with the president and a number of vice presidents, who had contributed to the company's success in several key new markets, and who was actively recruited to go overseas for two years:

> Your assignment is up, and you say, OK, now what job do I get when I go back? The answer is, go find one.
>
> It was very scary because I had been following what had happened to other expats that went back, and I knew of only one that actually came back to a good job, came back to a job at all. So I came back in December [on a brief trip] and I sent an e-mail to all the top people in the company and basically started trolling for jobs. The response was, "Great, come in, love to talk to you, let's see what might be available."

This manager considers himself lucky because a good job did open up while he was in the United States, and thanks to his connections at the top of the company, he was able to get it. But as he observed:

> If that spot hadn't been there, I would have come back from that trip with nothing. And I was facing my visa expiring in sixty days.
>
> I think it's a case of the company's not having the maturity as an organization to capture and leverage the investment that's been made in the people. People sent overseas on postings are not the bottom 10 percent of the company, they're the top 10 percent. And if we're making this kind of investment, it's an absolute sin that we're not recapturing this investment.
>
> But people come back at different points in time and try to reenter the organization, and there's no system in place to take advantage of the value the person has built up.

A business development manager working in Buenos Aires was three months from returning to the United States after a two-year posting. He was visited by the vice president for Latin America, who told him that because business in Japan was down, all salary and administration budgets in the international division were going to be cut for the third time, and that there would be no job for him when he returned. He would be given three months' severance pay when his assignment ended. He therefore spent most of the rest of his assignment trying to find another position with another company in the United States. This was difficult to do because he could not take time off to return on a job-hunting trip. In the end he did find a job, but the company that had sent him overseas had certainly lost his full attention to his job for a period of several months.

Taking Control

Well, is there anything you can do about this situation? Or is it totally bleak? Fortunately, some steps you can take can be very effective in setting yourself up for a successful reentry into your own company.

1. **Build the right relationships.** When you're back in the States for product training or for whatever reason, start your reentry job network, even as early as a year before you are due to return. You can rely on trusted friends to pass you information about reorganizations, new business directions, and other signs that new job possibilities are going to open up.

2. **Make friends with the recruiters.** Your company either uses in-house recruiters or an outside service. An individual recruiter usually has responsibility for a particular part of the business. Get acquainted with the recruiter responsible for your field. If she drinks Chardonnay, bring her a gift from that off-site meeting in Napa Valley. She will remember you and is in a position to keep you apprised of every opening that will be coming up, as well as the many political nuances surrounding the hiring process for each job, and who your competition is.

3. **Use the Internet.** Your company has a Web site, and all, or most of, the positions that are open are listed on it. You can get access to this list from anywhere in the world—something that was not possible when the people whose stories were told above were getting ready to return. You want to do this as a complement to your main sources of information—your friends and your favorite recruiters.

4. **Make an early trip back home.** If necessary, either arrange to have a business trip home a couple of months before your tour is up or buy the air ticket yourself. Set up your meetings as much in advance as possible, using your network.

5. **Be smart.** If you need to, you can accept a lateral transfer to get back home, into the company; then you can maneuver to get the job you really deserve. This is where the unspoken threat of leaving for another firm can be effective, but you'll have to move fast and determine within a few weeks of your return whether you want to stay with your old company or pull up stakes and move. Any potential new employer will understand why you want to look for a better job if you have put in two or three years of honorable service overseas, and you're now stuck in a job that does nothing to advance your career.

In some ways, working in an overseas high-tech assignment is like working in a start-up: You generally have much more authority and responsibility than you did back in the States, and you're part of a smaller group that can become quite cohesive over time. Working in a foreign environment can make you more sensitive to diversity, and more tolerant. It can awaken you to some of the shortcomings others may see in your own culture, and it can make you more aware and respectful of other cultures. My personal feeling is that you generally go back home a better person.

The high-tech industry needs foreign markets, and should you decide to continue in an international career, your overseas experience will make you very much more attractive as a potential hire to other companies, provided you work it so that you approach them while you're still employed by your present company.

And if you decide to return to the United States for good, your life will be richer in ways you can appreciate only when you have been through the overseas experience.

10

FINAL WORDS OF WISDOM

A career in high tech can be a wonderful, invigorating, rewarding occupation. Not many people find it easy to leave the high-tech industry for something else. It's still a very young industry, and there's something thrilling to being involved with technology that is in one sense so advanced and, in another, still in its infancy. Perhaps I'm just being an optimist, but I do feel that high tech has tremendous potential for good, and I like being a part of its development for that reason. Technology can be used for good or for evil, of course, but I like to think my eyes are wide open to some of the inane aspects of the high-tech world—for example, the belief that virtual reality is better, somehow, than the real world. The stuff does grow on you—no question about it.

By now you know that high tech pays well, offers all sorts of challenges, and is open to nontechnical people. You know there are good work situations and not-so-good ones, as with any work situation. But you also know that high tech offers involvement in a unique culture, and that it's less concerned with trivial things—race, dress, sexual preference, and so on—than with performance. I hope it sounds like a pretty good field for a first, second, or third career.

I'd like to touch briefly on three issues concerning high-tech careers that represent the less-attractive elements of the industry. First, it's too easy to lose yourself in the work. High tech seems to call on people's reserves of strength and energy, and they willingly give in. High-tech companies are usually understaffed, so there's too much to do. That doesn't stop individuals from trying to do it all. As you climb the ladder of promotions, the situation gets more demanding, not less. Managers work harder, directors work even harder, and vice presidents work harder than that. They get detached from the "real" world. That's a big price to pay, in my opinion.

The rewards are there, that's certain—pay, bonuses, stock options. If that's what is most important to you, fine, but many, many marriages and relationships have melted in the heat of the high-tech furnace, and children become estranged from parents. There's a very successful man I know who is estranged from his three daughters. He showered them with gifts and gave them expensive educations in the finest schools. They're now in their twenties, and he missed their growing up. He's sorry, deep down, but what can he do? The fortunes of a large enterprise employing tens of thousands of people rest mostly on his shoulders—or at least that's what he thinks. He couldn't refuse the responsibilities and rewards heaped on him, or betray the trust of the other top managers in the company, so the company is now his life.

Sure, I know that's true of careers on Wall Street, and maybe in the toothpaste industry, for all I know, but at least be aware of the danger because high tech doesn't just entrap the top people. Many middle managers and individual contributors also sign their lives over to high tech.

Another danger: Because of the fast pace, lack of structure, and inexperience of managers in high tech, feedback on your job performance tends to be unreliable. If you are the sort of person who needs confirmation that you are doing a great job, you may be disappointed, and even demoralized. High tech tends to leave its practitioners alone and without guidance. This is a positive element for some people, and a negative for others. Yes, there's a formal process for evaluating performance, but it's hard to evaluate performance when there are no meaningful benchmarks established or when you stray from what the manager, months later, feels you *should* have done. In short, you'd better be a well-grounded individual, accustomed to judging your own performance—or have some friends whom you can ask for reliable feedback. Do not trust the formal review system. High-tech managers are, generally speaking, not well trained in assessing employee performance and in giving helpful feedback. Also, they are usually too busy to give performance reviews much of the time. Companies instead rely on having managers and employees fill out forms, in an attempt to force some uniformity, at least, into the process. Often employees are asked by managers to write their own assessments, which the manager will then edit. A few years ago a manager confessed to me that he was simply going to take the reviews he had given for me the previous year, rearrange them a bit, and use them for the current year.

So save those congratulatory e-mails from those you've helped in a special file. They'll come in handy when review time rolls around. Keep a record of every major activity you've carried out as well. This documentation can be summarized for your contribution to your review, and you'll have full backup for everything you write.

Finally, high-tech careers get more risky—and more tiring—the older you get. If you are, say, fifty years old and making really good money, your career may be in danger. Chapter 7 gave you some good ideas for bombproofing your job. The chief thing you must do is maintain your network of friends, especially within the company. They will watch out for you and your interests.

But the demands of high tech for constant high performance, overwork, and continual reeducation can also tire you out. As Chris Toal, a fifty-year-old businessman reported in the *Wall Street Journal*, "You are proving yourself every day. I know I've peaked in my career. I am never going to be a vice president. But I'm running as fast as I can, and I don't know why."

High tech doesn't place any importance on the wisdom or insights of older workers or managers; everything is geared to performance at a frenetic pace. Some people, including older employees, thrive on this. Others do not.

It is, I believe, important to have a plan B in mind for those later years. If you're over fifty, and you've spent five years or more in high tech, you probably know enough to establish a small, profitable business of your own, providing services to high-tech companies. If you haven't given any thought to this—what your product is, who your initial customers will be, how you'll market yourself, and how much you'll charge—then the transition may be very difficult. So start preparing before you get close to that half-century mark. The skills and wisdom you've accumulated can be of great value to a young company when they are made available in the form of consulting.

I've solicited the opinion of friends who are in their fifties and sixties, who are working in high tech. Some of them even entered high tech in their fifties. They all agree that having alternatives makes sense, and most of them have given it serious thought. For some of them, though, their alternative plan doesn't involve high tech at all. One sales manager wants to open a marine supplies shop, and another wants to be a guide for wilderness treks. And why not, if the kids are out of the nest, and the house is paid off—or nearly paid off.

On balance, I can't think of any career that offers more to people like us than high tech. With all the risks and downsides, stresses and strains, it's provided me with a great deal of satisfaction, and I think it can do the same for you. I haven't let high tech devour my life, nor has it become my life. I've got some great friends who don't hesitate to let me know if I screw up—but who also give me nice feedback when things go well. I've had a plan B for a few years now, but haven't put it into operation yet. In the meantime, I'm slogging along with my colleagues in the industry, watching technology develop at speeds and in directions that are simply amazing. It's fun to come home, water the flowers on our deck, and try to do some serious reading. Hiking in the Sierras, thinking about stuff, and writing the occasional book are also pastimes that bring some balance into my life.

When you join the world of high tech, I believe that you, too, will face questions of balancing your "outside" life with fascinating, challenging work. Being a "low-tech" person, you're pretty well equipped to consider these matters from all angles. I wish you happiness in your career, and a rich and wonderful life.

Visit us at www.hightechcareers.net!

WORKS CITED

pages 12–13
"several years from now . . .": "What to Do About Microsoft," *Business Week,* April 20,1998.

page 13
Schonfeld, Erik, "The Network in Your House," *Fortune,* August 3, 1998.

page 87
"the first criterion is . . .": Peter Vozas as quoted in the *San Jose Mercury News,* June 14,1998.

page 92
"On the Hiring Line," *San Jose Mercury News,* July 19, 1998.

page 142
"A Java in Every Pot," *Business Week,* July 27, 1998.

page 154
Old technical manual for Sun Microsystems: *Multithreading and Real-Time in Solaris,* Sun Microsystems Inc.,1992.

Page 190
"Stress and emotional exhaustion . . .": Christina Maslach, "Burnout Gaining Recognition As an Employer Problem": *San Jose Mercury News,* December 26, 1997.

page 195
"the ability to actively manage . . .": Betsy Collard, "Creating a Resilient Workforce," a presentation given at a symposium, June 29, 1993.

page 199
"Is It Time to Start Bragging About Yourself?," *Fortune,* October 27, 1997.

page 248
"you are proving yourself . . .": Chris Toal, "Gray Expectations," *Wall Street Journal,* May 5 1997.

INDEX

access, 156
account management, 23–24
administrator, 24–27
ads, job. *See* classified ads
advertising, 53, 59, 146, 227–228
age and high-tech jobs, 188–189, 248–249
angel, 209
animation, 142
Apple Computer, 63–64, 139, 142, 166, 167, 208
applications (apps), 136
    compute-intensive, 137
    incompatible, 148
appraising yourself, 83–85. *See also* qualities needed; skills
architect, 156
architectures, computer, 138
area studies majors, 233
art history majors, 58
artists, 77–80, 146
ASIC (application specific integrated circuit), 20
attitudes
    adjusting your, 85–87
    desirable for high tech, 123
authority. *See* responsibility, assuming

batch mode processing, 136
behavioral interviewing, 125
benefits administration, 49
beta version, 134, 156
biology majors, 61
bit, 151
board of directors, 209
bonuses, 17, 247
bosses, 178
    bad, 189
    multiple, 24
    perfectionist, 176–177
box, 156
brainpower and high tech, 169
brand management, 52–53
branding, 53, 55
bugs, software, 148
bundle, 156
bureaucracy, freedom from, 212–213
burnout, 8–9, 189–190

bus, computer, 153
business development, 22, 27, 221
business letters, 185
business majors, 40–41, 46
business plan, 209
business skills, 35
Business Week, 133
byte, 151

Cabletron, 177
cabling, computer, 144, 162
cache, 153
California Psychological Inventory (CPI), 83–84
    call centers, 39
Campbell Interest and Skill Survey (CISS), 84
Career Action Center (Cupertino, CA), 182
career assessment, 84
career self-reliance, 180–184, 195
careers in high tech
    choosing, 5–11
    for low-tech people, 18–80
    starting, 81–130
    thriving in, 188–207
cars, computers in, 143
chalk talks, 185
channels, 156
channels development, 34
channels management, 2, 28–35, 221
channels strategy, 34
chaotic interview, 123
CISC (complex instruction set computing), 152
Cisco Corporation, 13
classes, taking, 178–180, 205–206
classified ads, 1, 6, 99–102. *See also* job listings on the Internet
client-server computing, 140
Clinical Practice of Career Assessment (Lowman), 84
clock speeds, computer, 152–153
cold approaches to job seeking, 96–98
Collard, Betsey, 182
collateral material, 61
college degrees and high-tech jobs, 1–2
3COM Corporation, 142

251

communication, informal styles of, 184
communication skills, 2, 5, 7, 42, 52, 66, 67, 72, 75, 90
   articulating ideas, 60–61
   employee communications, 48
companies. *See* high-tech companies
Compaq, 148, 166, 170
compensation. *See* bonuses; salaries
compensation administration, 49
compiler, 149
compute-intensive, 137
computer bus, 153
Computer Glossary: The Complete Illustrated Dictionary (Freedman), 89, 155
computer skills, 88
computer systems, 161
computers
   buying, 133
   and the home, 12–13
   learning about, 91
   overview of, 131–163
Computerworld, 88–89
configure, 156
consulting, 248
contacts, building, 27
continuing education, 179, 205–206. *See also* learning on the job
contract administration, 35–36
contract negotiation, 36–37
contract work, 5, 105–108
convergence, 13
cooperative processing (computer), 139–143
corporate communications manager, 37–38
corporate headquarters, 14
corporate marketing, 227
cost centers, 43–44
course developer, 44
courses, taking, 178–180, 205–206
cover letters, 113–117, 119
creative skills, 77–80, 146
credibility, 191
crisis management, 59, 60
CS. *See* customer service
cube, 156
cultural differences, 238–240
culture
   high-tech, 4–5, 11, 15–17, 164–187
   living in a foreign, 219
customer relations, 23–24
customer response center, 38–40
customer service, 38, 156, 222
customer service business manager, 40–41
customer service representative, 41–44
customer training, 44–45
customer visits, arranging, 25

customers
   types of, 222
   working with, 27, 38–40, 65, 70–71, 74–75

dance and high tech, 78
Data General, 177
data processing, 136
datapoint, 156
DEC, 137, 148, 166, 167, 170, 181
decentralized computing, 137–138
dedicated salespersons, 31
degrees, college, 1–2
dejobbing of America, 192
Dell Computer Corporation, 64
Delphi Automotive Systems, 142
developers. *See* course developer; software developers
devices, input and output, 153
devices, smart, 12–13, 142–143
Digital Equipment Corporation, 137, 148, 166, 167, 170, 181
direct sales, 29
directories, using effectively, 14, 97
distributed computing, 137
documentation, creating, 74
documents, careers dealing with, 37
doorway meetings, 184
down, 156
downloading, 140
downsizing, 156, 181, 183, 188
drag-and-drop, 157
dream jobs, 109–110
dress codes, 165, 171–172

e-mail, place in high tech, 185
earnings. *See* bonuses; salaries
educational backgrounds, 1–2, 86. *See also* names of specific fields, e.g. English majors
egalitarianism, 17, 170–172
electronic commerce, 146
embedded systems, 12–13
employee communications, 48
employee nurturance, 180–182
employee recruitment by firms, 6–11
employee referrals, 92
employee training and development, 48–49
employers, prospective
   researching, 125–126
   telephoning and visiting, 101–102
end user, 157
English majors, 2, 73, 74, 77
entry level positions, 82, 130
   international, 222, 229–230
   to take and not take, 104–105, 211–212

environment, 149. *See also* culture
EOL (end of life), 68–69, 157
ethernet, 143–144
ethnic diversity, 165
event planning, 33
events management, 53–54
expats, 225, 237
experience, acquiring, 78, 87–91, 115
extranets, 145

fashion design majors, 20
FCS (first customer shipment), 60, 69
FDDI (Fiber Distributed Data Interface), 144
fiber optic cable, 144
field marketing, 54–55, 227
film and high tech, 78
finance, jobs in, 58, 229
finance majors, 3–4
fine arts majors, 55
first high-tech job. *See* entry level positions
fit between candidate and employer, 110–111, 121
flattened corporation, 173, 181
floppy disks, 139
focal, 157
footprint, 138
forecasting (orders and shipments), 45–46, 56
foreign language skills, 3, 79, 219, 231–233
foreign studies majors, 231–233
Fortune, 133
foundry, 157
Freedman, Alan, 89, 155
FUD (fear, uncertainty, and doubt), 29
fun at work, 164–165
functionality, 157

Gates, Bill, 208
geo, 30, 157
gig, 157
goal, 157
going forward, 157
goodness, 158
government employment experience, 94
granular, 158
graphic art skills, 20, 77
groupware, 143
growing a job, 27, 174–175, 189, 191–198, 230–231. *See also* lateral transfers; moving up
GUI (graphical user interface), 149, 158

hallway meetings, 184
hard copy, 136
hard disks, 139
hardware companies, 13, 225–226

headcount, 158
headhunters, 7
help wanted ads. *See* classified ads
Hertz, Heinrich Rudolph, 153
heterogeneous networks, 150
Hewlett-Packard, 208
high tech, 12–17
high-tech companies
   culture, 15–17, 164–187
   differences from other companies, 15–17
   directories, 14
   employee recruitment by, 6–11
   failures, 213–214, 217
   locations, 14
   objectives, 165–166
   overseas activities, 225–229
   reorganizations, 190
   structure, 14–15
   types of, 13
high-tech industry
   changeability of, 16, 169
   characteristics of jobs in, 85
   less attractive elements of, 246–249
   related industries, 13
high-tech job descriptions, 18–80
   in ads, 1
hiring from inside, 129
hiring process, 127–130
   criteria, 99, 110–111, 116
history majors, 19, 61
homes, use of computers in, 12–13
hours of work, 177
HR. *See* human resources
HTML coding skills, 4, 78, 79
human resources, 46–49, 108, 229
   going through to get a job, 111–112, 114, 128
humanities majors, 74

IBM, 144, 166–167, 180–181
icon, 158
IEEE (Institute of Electrical and Electronic Engineers), 144
income. *See* salaries
India, 226
informality, 170–172, 184–186
information, organizing, 78
informational interviews, 26, 90, 114, 126
input devices, 153
inside reps, 66
insider referrals, 92
instruction set, 152
Intel Corporation, 139, 203
intelligence. *See* thinking skills
interface, 158

international experience, usefulness of, 79
international work
    advantages of, 218–220
    based in the U.S., 222–224
    considerations before doing, 238–240
    kinds of positions available, 220–222, 224–229
    obtaining, 224–225, 229–238
    returning to U.S. from, 196–197, 240–241
Internet, 145–146
    getting on, 133–134
    job listings on, 98–99, 244
    learning technology from, 89, 133
Internet-related jobs, 77–80
Internet Service Providers, 13
Internet start-up companies, 214–215
interning, 78, 115
interpersonal skills. *See* relationship skills
interviews, job, 81, 89–90. *See also* informational interviews
    courses on, 89–90
    handling, 121–127
    importance of, 110–111
    question to answer, 124
    questions to ask, 123, 187
intranets, 77, 140, 143, 145
Intuit, 64
investment banks, 210
iron, 158
ISV (independent software vendor), 25, 158
    working with, 49–50

jargon, 89, 132, 134–135, 154–163
Java, 142, 150
Jini, 64
job acceptance letters, 96
job ads. *See* classified ads
job autonomy, 174, 195
job candidates, successful, 110–111, 116, 121
job changing, 86, 188–189, 189, 196, 248–249
job descriptions, 18–80
    in newspaper ads, 1
job evaluations, 247–248
job experience, helpful prior, 24
job fairs, 108
job-focused interview, 125–127
job hunting, 81–130
    psychological hurdles to, 7–8
    search strategy, 82–83
    in start-ups, 214–216
    summary of steps in, 130
job listings on the Internet, 98–99
job referral bonuses, 92
job requirements, circumventing, 99
job satisfaction, 6, 16. *See also* opportunities

job security, 7–8, 168
job variety, 212, 216
jobs
    first high tech. *See* entry level positions; job descriptions
    getting stuck or stereotyped in, 189
    joint ventures overseas, 226
journalism majors, 55

LAN (local area network), 143–144
language skills, 3, 37, 163
    foreign languages, 79, 219, 231–233
large account management, 23–24
lateral transfers, 179, 203–205, 244
law-related jobs, 36–37, 103, 229
layoffs, 181
    effects on careers, 188–189, 214, 217
learning on the job, 21, 39, 43, 87, 178–180, 183. *See also* continuing education; reading about high tech
legacy, 139
legal-area jobs, 36, 37, 103, 229
letters
    asking for a job, 96
    business, 24–25
    cover, 113–117, 119
    job offer and acceptance, 96
leverage, 158
lingo. *See* jargon
listening skills, 35, 90
living overseas, 219, 221, 236, 238
    company expenditures on employees, 225
local hire, working as a, 236, 238
look and feel, 158
Lowman, Rodney L., 84

magazines about high tech, 88, 133
mail, delivering interoffice, 25
mainframes, 136
major account management, 23-24
management
    high-tech styles of, 173–178
    of teams, 32–33
managerial skills, 10
manufacturing operations, 225–226
map, 158
marcom, 15, 55
market development. *See* business development
marketing, 15, 50–56, 58, 59, 75, 197–198, 227–228
Maslach, Christina, 190
master distributors, 30–31
materials management, 56
mathematics skills, 57

McNealy, Scott, 11, 141, 172
media, 159
media relations, 37, 59–61
meetings, 25, 184, 186
megahertz (Mhz), 153
memory, computer, 140
memory modules, 153
memos, place in high tech, 185
mentors, 202–203, 240. *See also* networking
mergers, 190
micromanagers, 176–177
microprocessors, 150–154
Microsoft Corporation, 64, 139, 142, 167, 208
microwave transmissions, 145
middle management, 174, 179, 181
migrate, 159
migration path, 159
mindset, 87
mindshare, 159
minicomputers, 137–138, 139
MIPS, 159
MIS (management information system), 137
mission critical, 159
mission statements, 165–166
Mobile Oil, 181
motivated persons, 11, 75, 105
Motorola, 142
moving up, 58, 75, 82. *See also* growing a job quickly, 216
multimedia, 78, 79, 141
multiprocessing, 152
multitasking, 159
  computer, 20, 140, 147
  human, 40, 57, 61
music majors, 107
music-related skills, 10, 77
My Yahoo!, 89
Myers-Briggs Type Indicator, 84

national account management, 23–24
National Semiconductor, 176
negotiating skills, 35, 36–37
Netscape Communications, 203
Network (big dog), 228
networking, 3–4, 5, 81, 91–96, 111–112
  to get a job overseas, 234
  to get job in U.S., 243
networks, computer, 13, 140, 143–146. *See also* Internet; intranets
newspaper ads. *See* classified ads
Newton, 142
numbers, 159
nurturance culture, 180–182

OEM (Original Equipment Manufacturer), 31, 159
off-sites, 165
offer and acceptance letters, 96
O'Hara, Shelley, 133
older employees, 188–189, 248–249
Olsen, Ken, 170
open vs. proprietary systems, 140
operating systems, 138, 139, 147–150
operations, 56–58, 229
opportunities
  for finding high-tech jobs, 4, 14, 91–110, 146
  for newcomers to work overseas, 230–233
  while in high-tech jobs, 5, 8, 16, 25, 73, 188, 218–220. *See also* learning on the job
order operations (order ops), 58
organizational skills, 62, 65, 66, 78
OTE (on target earnings), 159
outplace, 160
output devices, 153
outside reps, 66, 75
overseas jobs, 221, 224–225, 238–240
overseas sales manager, 234

package, 160
partnerships, corporate, 71
pay. *See* salaries
PDA (personal digital assistant), 142
penetration, 160
people skills. *See* relationship skills
performance reviews, 247–248
peripherals, 13, 136, 147
persistence, 68, 87, 123
personal computers, 139
personality tests, 84
persuasiveness, 60, 72
philosophy majors, 78, 103
phone calls. *See* telephone calls
photography and high tech, 78
platform skills, 44
platforms, computer, 50
political science majors, 61, 103, 193
port, 50, 149
power supply, 153
PR. *See* public relations
presentations, 25, 185, 198–202
press relations, 59–61
pressure. *See* stress
processing speed, 152–154
procurement, 56
product distribution, 2
product knowledge, 15–16, 19, 30, 37, 52
product management, 20, 58, 61–65

product marketing, 61–65
products
  distribution, 56
  forecasting, 45–46
  in-house group sales of, 29
  life cycles, 141
  promotion, 59, 227
  transitions from old to new, 24
  warranty management, 35
professional positions, 18
profit centers, 43–44
programmers, 137
project coordinator, 65–66
promotions. *See* moving up
proprietary technology, 140, 148, 210
psychological tests, 83–84
psychology majors, 51
public relations, 19, 37, 59–61, 228
public sector experience, 94
public speaking. *See* presentations
publications
  high-tech company, 25, 60–61, 74, 227
  high-tech industry, 13
  to read, 84, 88–89, 133, 155
pushiness, 123

qualities needed. *See also* skills
  to go overseas, 235
  for working in high tech, 11, 85, 123
  for working in start-ups, 38, 67, 213–213, 216
Quicken software (Intuit), 64

RAM (random access memory), 140
reading about high tech, 15–16, 206
  to gain skills, 88–89
  recommended publications, 84, 88–89, 133, 155
real-time, 134, 153
recompiling, 149
recruiters, 99, 108, 244
Red Herring, 88-89
reentry to U.S., 196–197, 240–245
regional offices overseas, 229
relationship skills, 23, 31, 35, 37, 40, 41, 42, 54, 60, 72, 75
relationships, effect of job on, 247
reorganizations (reorgs), 160, 190
reprofile, 160
req, 160
research and development, 14–15, 18
responsibility, assuming, 216, 219, 244–245
resume
  broadcasting, 96
  compiling, 110–113, 118, 120
  international experience on, 237
rich, 160
rightsizing, 160
RISC (reduced instruction set computing), 152
risk capital, 209
risk-taking, 7–8, 166–169
robust, 160
rollout, 161
ROW (rest of the world), 238
RTFM (read the f____g manual), 161
RTU (right to use), 161
Russia, 226

salaries, 4, 9–10, 17, 21, 247. *See also* bonuses; compensation administration
  jobs with higher, 40, 66–67
  at start-up companies, 215–216
sales, 31, 58, 66–69, 74–75, 226–227
sales channels, 28
sales offices, 14
sales reps, 23–24, 27, 66
sales support, 70–71
satellites, 145
scripted interview, 124
SCSI, 161
search engines, 13, 133
secretaries vs. admins, 24
security, job, 7–8, 168
self appraisals, 83–85. *See also* qualities needed; skills
self-motivated persons, 11, 75, 105
self-reliance, 180–184, 195
semiconductor companies, 175–176
servers, 140
services (computer), 140
shipping department, 56
shirts, promotional, 165
Sixteen Personality Factor Questionnaire (16PF), 84
skills. *See also* qualities needed
  articulating ideas, 60–61
  business, 35
  calmness during crises, 61
  communication, 2, 5, 7, 42, 52, 66, 67, 72, 75, 90
  computer, 88
  creative, 77–80, 146
  English language, 163
  foreign language, 3, 79, 219, 231–233
  graphic art, 20, 77
  HTML coding, 4, 78, 79
  language, 37
  listening, 35, 90
  managerial, 10
  mathematics, 57

multitasking, 40, 57, 61
music-related, 10, 77
negotiating, 35, 36–37
organizational, 62, 65, 66, 78
persistence, 68, 87, 123
persuasiveness, 60, 72
platform, 44
pushiness, 123
relationship, 23, 31, 35, 37, 40, 41, 42, 54, 60, 72, 75
reliability, 54
spreadsheet, 88
teaching, 44, 76–77
technical, 42, 75
thinking, 34, 57, 78
time management, 40
word processing, 88
writing, 2, 37–38, 59–60, 72–74, 77, 185
smart devices, 12–13, 142–143
smart people, 169
sociology majors, 77
software, 148, 149, 162. *See also* applications (apps); operating systems
software companies, 13, 16, 226
software developers, 49–50, 149, 226
software industry, 101
Solaris, 150
Spanish language skills, 3
spiff, 161
sponsor for overseas employees, 240
spot bonus, 17
spreadsheet skills, 88
start-ups, working for, 38, 67, 208–217
status of admin jobs, 26–27
stock options, 215, 247
stocks, start-up company, 210
storage capacity, computer, 140, 147
strategic alliance manager, 71–72
stress, 190
   in start-ups, 216–217
Sun Microsystems, 11, 64, 141, 142, 150, 172, 177, 203, 228
supplier management, 56
supplies, ordering, 25
support, 161
support functions, 15
system (computer), 161
systems integrators, 32–33

taxes and overseas jobs, 236
TCO (total cost of ownership), 138
teaching skills, 44, 76–77
teaching vs. work in high tech, 73
teams, working in, 32–33, 65–66, 72, 74, 165, 174

technical editor, 72–73, 195
technical education, 76–77
technical skills, 42
   vs. aptitude for sales, 75
technical vs. non-technical jobs, 20
technical writer, 37–38, 73–74
technology
   learning about, 43, 131–136
   overcoming fear of, 3
telecommuting, 173
telephone calls
   call center jobs answering, 39, 40
   high-tech styles of answering, 184
   to prospective employers, 101, 114
telesales, 74–75
temp work, 78, 102–105
10 Minute Guide to PC Computing, 133
terminals, 137
terminology. *See* jargon
tests, psychological, 83–84
thinking skills, 34, 57, 78, 169
time management skills, 40
time spent at work, 172–173, 177
timeframe, 161
timeout, 162
Toal, Chris, 248
token-ring networks, 144
total cost of ownership, 138
trade shows, 25, 53, 56, 227
training, getting. *See* continuing education; learning on the job
training others, 76–77
   customers, 44–45
   employees, 48–49
   sales channels, 34–35
traits. *See also* qualities needed; skills
travel, jobs involving, 24, 27, 36, 37, 54, 60, 221, 244. *See also* international jobs
   infrequent trips, 222
twisted pair, 162

UNIX, 149–150, 155
up and running, 162
upgrades, 148, 162
user groups, 89, 133

value add, 162
value propositions, 34
values. *See* high-tech culture
VAR (value added reseller), 33–34, 162
VC (venture capitalist), 208–209
venture capital, 208–210
verbal communications in high tech, 185
version, 162
virtual, 162

VMS, 148
voicemail, place in high tech, 185
Vozas, Peter, 87

WAN (wide area network), 144–145
want ads. *See* classified ads
Web-based start-up companies, 213–215
web designer, 77–80
Web sites
   with animation, 142
   the author's, 249
   with job listings, 98–99
windows, 140
wireless technology, 142–142
word processing skills, 88
work pace in high tech, 172–173
   effect on older workers, 248
   pace in start-ups, 216
work schedules, 177
workaholism, 172–173, 177, 246–249
working abroad. *See* overseas jobs
workstations, 90–91, 139–143
writing skills, 2, 37–38, 59–60, 72–74, 77, 185
written communications, place in high tech, 185
WYSIWYG (wizywig), 163